COGNITIVE ONTOLOGY

The search for the "furniture of the mind" has acquired added impetus with the rise of new technologies to study the brain and identify its main structures and processes. Philosophers and scientists are increasingly concerned with understanding the ways in which psychological functions relate to brain structures. Meanwhile, the taxonomic practices of cognitive scientists are coming under increased scrutiny, as researchers ask which of them identify the real kinds of cognition and which are mere vestiges of folk psychology. Muhammad Ali Khalidi presents a naturalistic account of "real kinds" to validate some central taxonomic categories in the cognitive domain, including concepts, episodic memory, innateness, domain specificity, and cognitive bias. He argues that cognitive kinds are often individuated relationally, with reference to the environment and etiology of the thinking subject, whereas neural kinds tend to be individuated intrinsically, resulting in crosscutting relationships among cognitive and neural categories.

MUHAMMAD ALI KHALIDI is Presidential Professor of Philosophy at the Graduate Center, City University of New York. His book *Natural Categories and Human Kinds* was published by Cambridge in 2013.

COGNITIVE ONTOLOGY

Taxonomic Practices in the Mind–Brain Sciences

MUHAMMAD ALI KHALIDI

Graduate Center, City University of New York

Shaftesbury Road, Cambridge CB2 8EA, United Kingdom

One Liberty Plaza, 20th Floor, New York, NY 10006, USA

477 Williamstown Road, Port Melbourne, VIC 3207, Australia

314–321, 3rd Floor, Plot 3, Splendor Forum, Jasola District Centre, New Delhi – 110025, India

103 Penang Road, #05–06/07, Visioncrest Commercial, Singapore 238467

Cambridge University Press is part of Cambridge University Press & Assessment, a department of the University of Cambridge.

We share the University's mission to contribute to society through the pursuit of education, learning and research at the highest international levels of excellence.

www.cambridge.org
Information on this title: www.cambridge.org/9781009223621

DOI: 10.1017/9781009223645

© Muhammad Ali Khalidi 2023

This publication is in copyright. Subject to statutory exception and to the provisions of relevant collective licensing agreements, no reproduction of any part may take place without the written permission of Cambridge University Press & Assessment.

First published 2023
First paperback edition 2024

A catalogue record for this publication is available from the British Library

Library of Congress Cataloging-in-Publication data
NAMES: Khalidi, Muhammad Ali (Professor of philosophy), author.
TITLE: Cognitive ontology : taxonomic practices in the mind–brain sciences / Professor Muhammad Ali Khalidi.
DESCRIPTION: New York, NY : Cambridge University Press, 2022. | Includes bibliographical references and index.
IDENTIFIERS: LCCN 2022025094 |
ISBN 9781009223669 (hardback) | ISBN 9781009223645 (ebook)
SUBJECTS: LCSH: Cognitive science. | Cognitive psychology. | Philosophy of mind.
CLASSIFICATION: LCC BF311 .K42 2022 | DDC 153–dc23/eng/20220824
LC record available at https://lccn.loc.gov/2022025094

ISBN   978-1-009-22366-9   Hardback
ISBN   978-1-009-22362-1   Paperback

Cambridge University Press & Assessment has no responsibility for the persistence or accuracy of URLs for external or third-party internet websites referred to in this publication and does not guarantee that any content on such websites is, or will remain, accurate or appropriate.

For Tarif

Contents

List of Figures — page ix
List of Contributors — x
Preface — xi

1 **Cognitive Kinds** — 1
   1.1 Introduction — 1
   1.2 Naturalism about Kinds — 3
   1.3 Ontological Matters — 8
   1.4 Reductionism — 13
   1.5 Realism — 24
   1.6 Conclusion — 31

2 **Concepts** — 33
   2.1 Introduction — 33
   2.2 Empirical Accounts: Cognitive Neuroscience — 37
   2.3 Empirical Accounts: Cognitive Psychology — 46
   2.4 A Functional Account of Concepts — 56
   2.5 Objections and Replies — 68
   2.6 Conclusion — 72

3 **Innateness** — 75
   3.1 Introduction — 75
   3.2 Critiques of the Innateness Category — 78
   3.3 Innateness as a Cluster Category — 79
   3.4 Is Innateness a Homeostatic Property Cluster? — 88
   3.5 Objections and Replies — 94
   3.6 Conclusion — 98

4 **Domain Specificity** — 100
   4.1 Introduction — 100
   4.2 Domain Specificity and Its Confounds — 101
   4.3 A Preliminary Example and a Theoretical Proposal — 104
   4.4 Further Evidence — 109

	4.5	A Theoretical Challenge and Response	115
	4.6	Conclusion	120
5	**Episodic Memory**	123
	5.1	Introduction	123
	5.2	What Is Episodic Memory?	127
	5.3	Empirical Challenges	137
	5.4	Episodic Memory as a Cognitive Kind	142
	5.5	Is Episodic Memory a Neural Kind?	150
	5.6	Conclusion	155
6	**Language-Thought Processes**	158
	6.1	Introduction	158
	6.2	Empirical Evidence for the LT Hypothesis	161
	6.3	Formulations of the LT Hypothesis	163
	6.4	Proposal and Discussion	170
	6.5	Two Kinds of LT Process	174
	6.6	Conclusion	179
7	**Cognitive Heuristics and Biases** *(cowritten with Joshua Mugg)*	181
	7.1	Introduction	181
	7.2	Heuristic as a Cognitive Kind	184
	7.3	Sub Categories of Heuristics as Cognitive Kinds	188
	7.4	Confirmation Bias or Myside Heuristic	199
	7.5	Conclusion	208
8	**Body Dysmorphic Disorder** *(cowritten with Amy MacKinnon)*	210
	8.1	Introduction	210
	8.2	Characterization of BDD	215
	8.3	Comparison of BDD with OCD	218
	8.4	Proposal for a Causal Model of BDD	221
	8.5	Objections and Replies	226
	8.6	Conclusion	229
9	**Epilogue**	231
	9.1	Introduction	231
	9.2	Etiological–Environmental Individuation	232
	9.3	Ontological Categories	237
	9.4	Reductionism and Cognitive Neuroscience	239

References   243
Index   268

Figures

1.1	Alternative to the "layer-cake" view	*page* 15
3.1	Causal network associated with the kind *innate cognitive capacity*	88
5.1	Two possible taxonomies of memory	125
5.2	Schematic diagram of *episodic memory* (capacity) and *episodic memory* (state)	148
6.1	Language-thought processes are not a cognitive kind	179
7.1	Comparison of taxonomies of heuristics and biases from five different sources	191
7.2	Tripartite model of the mind proposed by Stanovich (2010)	196
8.1	Causal model of *body dysmorphic disorder*	224
8.2	A cognitive behavioral model of body dysmorphic disorder proposed by Veale (2004)	227

Contributors

JOSHUA MUGG (coauthor, Chapter 7) is an assistant professor at Park University in Kansas City, where he coordinates philosophy, religion, and interdisciplinary studies. He works in the philosophy of psychology, mind, and religion, primarily studying cognitive architecture, belief, and faith.

AMY MACKINNON (coauthor, Chapter 8) is a PhD candidate at Western University studying the philosophy of psychiatry, mind–brain sciences, and disability.

Preface

If a sudden interest in taxonomy is indicative of turmoil in a scientific field, then the cognitive sciences may be in a current state of crisis. Psychologists, neuroscientists, and researchers in related disciplines have recently devoted increasing attention to the ways in which their respective disciplines classify and categorize their objects of study. They are especially interested in the precise manner in which the taxonomic categories of psychology relate, or will relate, to those of neuroscience (e.g. Price & Friston 2005; Poldrack, Halchenko & Hanson 2009; Anderson 2015). This interest in "cognitive ontology" can be traced in part to a range of surprising results emerging from recent neuroimaging research. Despite a widespread expectation that advances in neuroimaging would lead to the discovery of neat match-ups between neural structures and psychological or cognitive functions ("structure-function mapping"), the results of the past few decades have been somewhat mixed when it comes to identifying precise cognitive functions for specific brain regions or networks. This has led many researchers to revisit the question of the relationship between cognitive and neural taxonomy and has led to renewed attention to taxonomic practices more generally.

Broadly speaking, four taxonomic approaches can be discerned among researchers in the cognitive sciences when it comes to structure-function mapping. The first, and perhaps most intuitively plausible position, *localism*, holds that there will eventually be a one-to-one mapping from brain regions or neural networks to cognitive or psychological functions (e.g. Young & Saxe 2009; Saxe 2010). Localism need not be committed to the claim that each cognitive function is performed by a single brain region, since networks of regions may well be implicated in performing any given cognitive function. But it does seem committed to the idea that the very same brain region cannot perform two distinct cognitive functions in different circumstances or on different occasions. In other words, each neural region is "maximally sensitive" or "selective" for a cognitive function.

Another view, *globalism*, claims that the entire brain, or at any rate, large swathes of the cortex, are implicated in carrying out any given cognitive function (e.g. Crick & Koch 1990; Suppes, Han, Epelboim et al. 1999). These functions are executed by the brain as a whole, and what distinguishes them from one another are such properties as levels of activation. A third view, *revisionism*, has gained prominence recently. It claims that our current cognitive categories need to be revised before they can be directly associated with neural categories. Our existing cognitive taxonomy is simply not suitable for mapping onto neural taxonomy and must be modified before a mapping can be effected (e.g. Price & Friston 2005; Poldrack, Halchenko, & Hanson 2009). Finally, the fourth position, *contextualism*, holds that different brain regions perform different functions depending on context (e.g. McIntosh 2004; Anderson 2010; Lindquist, Wager, Kobel, et al. 2012; Klein 2012). According to one incarnation of this view, it is the *neural* context that determines which function any particular brain region performs. That is, the cognitive function of a region depends on the other regions with which it is coactive, and this relies, in turn, on the nature of the neural pathways that connect regions, their patterns of connectivity (e.g. feed-forward, feedback), the type of coupling that obtains between them, and the temporal dimensions of that coupling. It also depends on such factors as neuromodulation by genetic and chemical means (Anderson 2015). But there is also another version of contextualism, which assigns different cognitive functions to a brain structure depending not (just) on the neural context, but also on the broader *environmental* context, as well as on the *developmental context* and on *causal history* (cf. Pöyhönen 2015; Hutto Peeters, & Segundo-Ortin 2017). This book argues that this variant of contextualism has been unjustly neglected and deserves greater attention. One of its guiding hypotheses is that certain cognitive constructs lend themselves to contextual (or environmental) and etiological individuation, thereby obstructing a one-to-one structure-to-function mapping.

The contextualist position entails that there can be a many-to-many relationship between structure and function (Khalidi 2017; 2020). It is not merely the case that the very same cognitive functions can be subserved by different neural structures or mechanisms (multiple realization), but the very same neural structures can also subserve different functions depending on context (multifunctionality). This will result in a crosscutting relationship between taxonomic categories that track neural structure and those that track cognitive function. This should not be such a surprising result, since crosscutting structure-function taxonomies are also in evidence in other areas of science, particularly the biological sciences. It is not difficult to find cases in other scientific domains in which there exists a many-to-many mapping between structure and function (cf. Weiskopf 2011; Stinson 2016). The function of flight is performed by different types of structure in different animals, such as bat wings and dragonfly wings. Thus, flight is multiply realized in structures as diverse as mammalian forelimbs and insect exoskeletons. At the same time, the mammalian forelimb performs a different locomotive function in different species, namely flying in bats, walking in cats, and swimming in dolphins (cf. Ereshefsky 2012). However, one of my central claims in this book goes further than this, since I will argue that there can be a many-to-many mapping not just from *structure* to function, but from types of neural state to types of cognitive state in general. This obtains even when one takes into account nonstructural aspects of neural phenomena, including such phenomena as chemical neuromodulation of brain regions.

What is the evidence for a many-to-many mapping between mind and brain, or between cognitive and neural categories? Despite skepticism among some philosophers (Kim 1992; Shapiro 2004; Polger 2009), I would argue that multiple realization, or a *many-to-one* brain–mind mapping, is well attested in cognitive science (e.g. Aizawa & Gillett 2009; Aizawa 2017). Moreover, a *one-to-many* neural-to-cognitive mapping is also a serious possibility. One striking finding from the last decade or two of research in neuroscience, particularly neuroimaging studies, is the extent to which particular brain regions are implicated in a range of seemingly very distinct psychological processes. The idea of neural pluripotency or "neural reuse" has been defended on the basis of a growing body of empirical evidence (Anderson 2015). To take just one example, the amygdala is involved in processing fear and other negative emotions, but also perception of odor intensity, sexually arousing stimuli, trust from faces, biological motion, and sharp contours (cf. Nathan & Del Pinal 2016). Some researchers would continue to insist that the apparent multifunctionality of brain regions stems from an insufficiently fine-grained individuation of these regions and that closer inspection will reveal that it is not the very same brain region implicated in diverse cognitive functions (e.g. Scholz, Triantafyllou, Whitfield-Gabrieli, et al. 2009). But others are at least open to the possibility that brain regions are indeed multifunctional depending on context. Drawing on a number of case studies, this book will ascertain the extent to which many-to-many mappings exist across the cognitive sciences, and will investigate whether this is due (at least in part) to the contextual individuation of cognitive functions, particularly when it comes to the broader environmental–etiological context. One of the principal themes of the book is that cognitive phenomena are often individuated with reference to environmental and etiological factors, but neural phenomena are usually not so individuated. Even when they are, the pertinent "external" factors do not always coincide with those relevant to the cognitive phenomena. That is why there can be a many-to-many mapping between cognitive and neural categories. This is not just a claim about the causal relationship between mind and environment but an individuative claim about the taxonomic categories that pick out real kinds in the cognitive domain.

The search for cognitive kinds has occupied many philosophers of cognitive science, especially those who have looked closely at particular case studies in the cognitive domain. There would appear to be two countervailing impulses among contemporary naturalist philosophers. The first is the tendency on the part of some philosophers to admit more or less any items into our cognitive ontology that are countenanced by our best scientific theories, thereby inviting the accusation that they are too permissive and are over-populating our cognitive ontologies. This kind of "rainforest realism" (Ladyman & Ross 2007) is sometimes even thought not to be a brand of realism at all, since allowing in too many items can be seen to undermine the idea that there is a select or elite group of ontological entities that make up the world (including the mental world). The other tendency among naturalist philosophers is to subject empirical work to detailed and meticulous scrutiny and to conclude that the psychological or cognitive constructs in question are not, despite appearances, natural kinds, thus inviting the rebuttal that they are being too restrictive.[1] How should we adjudicate this apparent tension among cognitive ontologists? There is clearly no substitute for looking at each case and judging it on its merits – and there is obviously no inconsistency in considering some cognitive categories to correspond to natural kinds and others not. For most of the case studies that I will be discussing, I will conclude that there are

---

[1] Both tendencies can be illustrated by the titles of publications by empirically minded and naturalist researchers who view the cognitive sciences through the lens of natural kinds. For the restrictive tendency, consider: "Is *thinker* a natural kind?" (Churchland 1982); "Consciousness is not a natural kind" (van Brakel 1995); "Psychiatric disorders are not natural kinds" (Zachar 2000); "Concepts are not a natural kind" (Machery 2005); "Memory is not a natural kind" (Michaelian 2011); "Pain is not a natural kind" (Corns 2012); "Addiction is not a natural kind" (Pober 2013); "Is emotion a natural kind?" (Griffiths 2004). Meanwhile, the permissive tendency can be discerned from the following titles: "The natural kind status of emotion" (Charland 2002); "Delusion as a natural kind" (Samuels 2009a); "Why don't concepts constitute a natural kind?" (Samuels & Ferreira 2010); "Depression and suicide are natural kinds" (Tsou 2013); "Addiction-as-kind hypothesis" (Ylikoski & Pöyhönen 2015); "Innateness as a natural cognitive kind" (Khalidi 2016a); "Autism as a psychiatric kind" (Weiskopf 2017a).

good grounds for considering them to be natural kinds. However, there are a few caveats that come along with this conclusion. First, scientific theories are defeasible, so the final verdict on any scientific category awaits the end of inquiry and our current conclusions are inevitably just tentative. Still, wholesale revision of concepts is rare in the history of science (which may be why the infamous categories of *phlogiston* and *caloric* recur so frequently in philosophical discussions of these matters), so we should not expect our current cognitive categories to be summarily swept aside and replaced in their entirety. Second, even though I will argue for a positive verdict for most of the categories to be discussed, I will also occasionally point to other categories in the vicinity (e.g. some superordinate or subordinate categories) that do not seem to correspond to natural kinds. Thus, most of the focal cases I discuss will be vindicated, but some of their close relatives will not. Third, even when it comes to the focal cases, I will sometimes argue for splitting rather than lumping, making the case that what is often taken to be a single natural kind is likely to be two or more different kinds. Moreover, in some of these cases, I will also argue that the split kinds do not belong to an overarching kind that comprises all and only those kinds. Hence, this amounts to a revisionary approach to our taxonomic practices and requires us to make some adjustments to our cognitive ontology. Finally, one reason why my conclusion tends to be less revisionary than that of some other philosophers, who have cast doubt on the existence of some central items in our mental ontology (e.g. *concept, emotion, pain*), is that I am operating with a more expansive notion of natural (or real) kind. As I will explain in the first chapter, the conception of natural kind that I am deploying (and have argued for elsewhere) is a nonessentialist one that relaxes some of the conditions that other philosophers have put on natural kinds. Others (including some cognitive scientists) who have tackled this question have been relying on what I consider to be an unduly restrictive or reductionist notion of kinds.

Chapter 1 will set out the broad metaphysical picture that will guide the inquiry. I derive the naturalist notion of kinds that I am using from the nineteenth-century discussion of classification and kinds initiated by Whewell, Mill, and Venn, rather than the more recent essentialist view of natural kinds put forward by Kripke and Putnam. I go on to defend a "simple causal theory" of cognitive kinds (Craver 2009), which conceives of them as "nodes in causal networks" (Khalidi 2013; 2018) in the cognitive domain. In addition, I argue against the layer-cake picture of scientific domains associated with Oppenheim and Putnam (1958) and put forward some reasons to resist reductionism when it comes to cognitive categories, based on different bases for individuating cognitive and neural categories. Finally, I respond to some concerns that the resulting ontological picture is not a realist one, on the grounds that it countenances the existence of cognitive kinds that are mind-dependent and self-reflexive.

In Chapter 2, I take up one of the most basic putative cognitive kinds, *concept*, arguing that it should be considered a real kind based on our current state of knowledge, contrary to what some philosophers have urged (e.g. Machery 2009). After surveying a body of empirical work on concepts in both cognitive neuroscience and cognitive psychology, I try to show that this work is pitched at two or three different levels of explanation. Much of the recent work on concepts using neuroimaging techniques should not be expected to reveal the neural correlates of concepts. That is partly because the research has different explananda and is investigating different causal processes. Meanwhile, other work on concepts in cognitive science reveals psychological structures (prototypes) associated mainly with automatic processing rather than deliberative reasoning. By contrast, concepts proper can be understood as functional kinds, which are individuated partly etiologically and partly with reference to the thinker's discriminatory and inferential abilities. I argue that many research programs in cognitive science individuate concepts in this way, combining diachronic and synchronic factors, though this does not seem to have been widely noticed by philosophers or psychologists. The resulting account of concepts is closely related to the "wide functionalist" theories first proposed by Harman (1982) and Block (1986), and is pitched at what Marr (1982) would call the "computational level," rather than the "algorithimic" or "implementational" levels.

Chapter 3 is about the category of *innateness*, which is a feature often associated with a range of cognitive phenomena, including concepts, cognitive capacities, behavioral dispositions, and mental states. Arguing against a number of recent critiques of the notion (e.g. Griffiths 2002; Mameli & Bateson 2006), I try to show that innateness can be identified with a cluster of properties that are causally interrelated in various ways and will propose a tentative causal model of the kind. In individuating innateness, it is important to distinguish proximal from distal causation. Some of the causal properties associated with innateness are involved in individuating innate cognitive capacities synchronically, while others are etiological in nature, responsible for making those capacities innate in the first place. This complex causal network is robust enough to warrant considering innateness to be a real kind as used in contemporary cognitive science. (This chapter is closely based on previously published work [Khalidi 2016a].)

Chapter 4 considers a related cognitive construct, *domain specificity*, which is invoked in a number of different research programs in cognitive science, to indicate cognitive capacities that are limited in certain ways. Specifically, the idea is that some cognitive capacities are restricted in their application to a certain domain, whereas others range freely beyond that domain. The challenge arises in saying what constitutes the domain of a capacity, especially since areas of knowledge do not come antecedently compartmentalized. Building on the work of some cognitive scientists, I argue that the best way to understand the proper domain of a cognitive capacity is by invoking evolutionary considerations. This means that domain-specific capacities are individuated etiologically (at least in part), based on their evolutionary history. They are also identified on the basis of their synchronic causal powers, what they can and cannot do, since domain-specific cognitive capacities cannot range beyond their proper domains (whereas domain-general ones can). Given this cluster of causal features, I argue that there is a prima facie case to be made for considering domain specificity to be a cognitive kind, one that may include various types of cognitive capacity, such as alarm calls in vervet monkeys and face recognition in humans.

In Chapter 5, I discuss the kind *episodic memory*, which has recently garnered a great deal of attention from philosophers. In light of current empirical work, it has become increasingly challenging to accept an influential and intuitively plausible philosophical account of memory, namely the causal theory (Martin & Deutscher 1966). It is unlikely that each episodic memory can be associated with a trace or "engram" that can be shown to be linked by an uninterrupted causal chain to an episode in the thinker's past. Some philosophers and psychologists have responded by effectively abandoning the category of episodic memory and assimilating memory to imagination or hypothetical thinking (e.g. Suddendorf & Corballis 1997; Michaelian 2016; De Brigard 2014). But I will argue that there is still room for a distinct cognitive kind, *episodic memory*, a cognitive capacity whose function it is to generate representational states that are connected to past episodes in the experience of the thinker, bearing traces of these episodes that are individuated not at the neural level but at the "computational level" (Marr 1982).

Chapter 6 considers an unusual cognitive category, which pertains to a kind of *process* rather than a kind of entity, state, or capacity, namely what I call *language-thought processes*. The kind of process in question is often discussed in the cognitive science literature under the headings of "linguistic relativity" and "linguistic determinism." I claim that these labels aim to identify a distinctive type of cognitive process, all of whose instances share something important in common, namely a fundamental or deep-seated influence of language on thought. However, by looking at some paradigmatic cases, I argue that there is nothing to distinguish this type of process from a broader cognitive phenomenon, namely concept acquisition or conceptual change. Moreover, I also argue that within this broader category, there are two distinct kinds of process that are usually lumped together that do not seem to have anything significant in common. There is an important difference between those processes that involve simultaneous recruitment of linguistic capacities and those that do not. I argue that these two types of process may constitute distinct cognitive kinds within the broader cognitive kind of *concept acquisition* or *conceptual change*.

Chapter 7 discusses the categories of *cognitive heuristic* and *cognitive bias*. These categories have come to define a burgeoning research program in cognitive science (the "heuristics and biases" program) and are widely considered to be universal features of human thought. On closer inspection, both categories are found to be too heterogeneous to identify real cognitive kinds, though some of their sub categories may be such. In particular, the chapter examines the construct *myside heuristic* (closely related to the phenomenon often known as "confirmation bias"). This is found to be a better candidate for cognitive kindhood, since it seems to pertain to a specific feature of human cognitive architecture. Moreover, the myside heuristic, which (roughly speaking) attaches more weight to one's own opinions than to contrary opinions, can be rational in certain contexts. Thus, distinguishing the heuristic from a corresponding bias can only be done against the background of a cognitive task or problem. This constitutes yet another instance of contextual or environmental individuation of a cognitive construct. Again, this kind of contextualism does not preclude it being a real cognitive kind, but it does make it unlikely that it will correspond to a neural kind.

Chapter 8 tackles a psychiatric kind that does not pertain to cognitive science narrowly conceived, though it is strongly rooted in cognition. It concerns body dysmorphic disorder (BDD), a psychiatric disorder that has been classified in the most recent edition of the standard *Diagnostic and Statistical Manual of Mental Disorders* (DSM-5) as one of the obsessive compulsive–related disorders (OCRDs). This condition involves persistent and intrusive thoughts about a perceived bodily flaw that is not observable or appears slight to others, it leads to repetitive behaviors, and it tends to result in significant distress or functional impairment. The chapter argues that the disorder seems to have an important cognitive component involving certain deficits in visual processing, in interpreting the mental states of others, and in assessing evidence for and against one's beliefs. A causal model of BDD is proposed that aims to show how its main features fit together. Based on this causal model, there are strong grounds for considering it a distinct psychiatric kind. The causal model also strongly suggests that it should not be categorized with the OCRDs. This model suggests a revision of the standard psychiatric taxonomy based on an analysis of the underlying causes of the disorder as opposed to its superficial symptoms.

Finally, Chapter 9 is an epilogue that brings together some of the main themes that run through the previous chapters. The principal themes that I recap are the etiological–environmental individuation of cognitive kinds, the advantages of a real-kind approach to cognitive ontology, and the purview of cognitive neuroscience. On the first score, I distinguish the variety of externalism defended in this book from the familiar varieties in the philosophical literature. On the second, I show how taxonomic practices in cognitive science can benefit from reflecting on the overarching ontological categories in the cognitive domain and on greater clarity in distinguishing relationships among different kinds of kinds (e.g. subordinate and superordinate kinds). On the third point, I argue that the scientific discipline of cognitive neuroscience, which aims to build bridges between neural and cognitive taxonomies, need not revolve around the search for neural correlates of cognitive kinds. After all, a scientific discipline like ecological genetics does not seek genetic correlates of ecological constructs.

This account of cognitive ontology cannot (obviously) hope to be comprehensive. But by choosing a number of central or representative entities for investigation, I aim to establish a number of conclusions about real kinds in the cognitive domain. First, when approached using a notion of kinds that is naturalistic and sensitive to the character of the special sciences, the cognitive domain can be seen to be populated by real kinds, which can be revealed by scientific investigation. Second, many of the items posited in our current scientific taxonomies are likely to correspond to real kinds in the cognitive domain, and wholesale revisions to our ontology are not likely. Third, notwithstanding the previous claim, some of the cognitive entities discussed in the book, and in particular, some of the ways in which some cognitive entities are classified (within superordinate categories, or into subordinate categories) are likely not to be consecrated or borne out by future scientific theorizing. Fourth, the relationship between our cognitive and neural taxonomies is complicated by the fact that etiological and relational individuation in cognitive science may lead to widespread mismatches between cognitive and neural categories, due to different individuative practices in the respective sciences. Finally, it is possible to be a realist about cognitive ontology while at the same time admitting that there may be crosscutting classificatory practices within the cognitive domain, yielding crosscutting taxonomies that track orthogonal causal processes.

* * * * *

This book might not have seen the light of day without an Insight Grant from the Social Sciences and Humanities Research Council of Canada, for the project "Taxonomic Practices in the Mind-Brain Sciences." I am grateful to this research grant in no small part for encouraging me to collaborate with others in ways that I would certainly not have been able to do without it. Collaborative work is rare in philosophy not just because resources are scarce, but because of a traditional disciplinary ethos of solitary contemplation. Notwithstanding this prevalent attitude, having the funding to help support research assistance is one way to encourage collaboration between established and emerging scholars. Thanks to this grant, I had the resources to support two junior researchers while working on this book, Dylan Ludwig and Amy MacKinnon (both of whose contributions I will detail below).

My largest debt is to my two coauthors on Chapters 7 and 8, Joshua Mugg and Amy MacKinnon, respectively, who very kindly agreed to contribute their research and writing to a monograph that is largely the work of someone else. Over a decade ago, Josh joined the graduate program at York as a very enterprising MA student who was eager to work on foundational questions in metaphysics. He went on to pursue a thesis topic that combined metaphysics and cognitive science, under my supervision, on the theoretical underpinnings of dual-process (or dual-system) theory in the philosophy of psychology. While he was finishing his dissertation, we collaborated on a commentary for the journal *Behavioral and Brain Sciences* on a target article proposing a new cognitive heuristic (Cimpian & Salomon 2014). This led to further collaboration on the topic of cognitive heuristics and biases, resulting in an article, "Self-Reflexive Cognitive Bias," which was published in the *European Journal for the Philosophy of Science* (Mugg & Khalidi 2021). When it came time to write a chapter on cognitive heuristics and biases, I very naturally reached out to him (now an assistant professor at Park University) for possible collaboration, and was delighted when he agreed. My association with Amy was very fortuitous.

I first met her while she was an MA student at Western when I attended two conferences there. After a couple of engaging conversations about the philosophy of psychiatry, which was a budding interest of mine, we agreed to stay in touch. When she briefly joined the PhD program in Critical Disability Studies at York University, we renewed our conversations about the philosophy of psychiatry. She soon decided to return to philosophy and to Western to pursue a PhD in philosophy, but in the meantime, she agreed to work as a research assistant on this book project, helping me investigate a number of different topics, notably, episodic memory. But our main work together has been on the psychiatric condition of body dysmorphic disorder, about which we read and discussed dozens of research articles. Her knowledge of psychiatry and the philosophy of psychiatry, as well as her firsthand experience working with patients with mental health conditions, was invaluable in helping me think through the issues, both theoretical and practical, so I was delighted when she agreed to be a coauthor on Chapter 8.

Even though he is not listed as a coauthor on any of the chapters, Dylan Ludwig has left scarcely less of a mark on the book than my two collaborators, and his contribution has been as important. While he was a PhD student working with me at York, I employed Dylan as a research assistant on this project from its very inception, and his help has touched every chapter without exception. His astute analytic skills, his ability to link philosophy and the sciences, and his background in neuroscience all helped shape this book in numerous ways. He was an invaluable conversation partner and sounding board on every chapter. On a more mundane note, he also prepared the bibliography and index for this book, with model efficiency and attention to detail.

Conversations with my colleagues at York University, particularly Kristin Andrews, Jacob Beck, Brian Huss, Kevin Lande, Alice MacLachlan, Robert Myers, and Claudine Verheggen, sometimes on the Toronto streetcar or subway, were more influential on my thinking about these issues than they realize. In some cases, a stray remark or innocent-sounding challenge led me to rethink some of the basic assumptions I was making. They have also organized and participated in a number of stimulating workshops at York over the years, which were very thought-provoking and left a long-lasting impression.

In June 2019, Joshua Mugg and I organized a workshop at York University on "Natural Kinds in Cognitive Science," which was a source of inspiration for writing this book and helped launch me on this project. I am very grateful to all the participants for their enlightening presentations and many contributions to the discussion sessions: Sara Aronowitz, Dan Burnston, David Colaço, Javier Gomez Lavin, Dan Kelly, Dale Stevens, Jacqueline Sullivan, Stephen Setman, and Maggie Toplak. Lively discussions after each paper and an informal roundtable at the end brought together all participants with members of the audience. The workshop was partially supported by the SSHRC grant mentioned above, as well as by the Department of Philosophy, the office of the Vice President for Research and Innovation, and the office of the Provost. Also at York, I was fortunate to participate for around five years in regular lab meetings of the Cognitive Neuroscience lab, led by Shayna Rosenbaum. The lab's research on episodic memory and spatial navigation helped me better understand how cognitive scientists validate their constructs and how they think about the ontology of the cognitive sciences.

In the midst of writing this book, I was very fortunate to be offered a position at the Graduate Center, City University of New York, which has been an ideal environment for finishing it. The lighter teaching load helped me meet the projected deadline (give or take a couple of weeks) and the intensive engagement with insightful graduate students inspired me in the home stretch. The final stages of writing this book were also contemporaneous with attending (virtual) events at the Graduate Center, notably the Philosophy Department Colloquium and the Cognitive Science Speaker Series. In addition, I have been a keen member of the Experimental Philosophy lab, organized by Jesse Prinz, with an ever-expanding cast of perceptive characters working on various projects at the intersection of philosophy and cognitive science. All these discussions informed the final drafts of these chapters in tangible and intangible ways.

Some of these chapters have appeared or have been presented in earlier forms at various forums. A version of Chapter 3 was published as: Innateness as a natural cognitive kind. *Philosophical Psychology*, *29* (2016), 319–333. I am grateful to the editors of that journal for permission to reprint it here (with significant revisions). An early version of Chapter 3 was presented at the Society for Philosophy of Science in Practice, Toronto, June 2013, and at the European Society for Philosophy and Psychology, Granada, Spain, July 2013. A much earlier version of Chapter 4 was presented at the Cognitive Science Society Conference, Portland, August 2010. Chapter 5 was influenced by a graduate seminar on Memory that I taught at CUNY in the Spring semester of 2020–2021, and the enlightening discussions with the students in that class. A version of Chapter 6 was also presented at the European Society for Philosophy and Psychology, Granada, Spain, July 2013. Finally, versions of Chapter 8 were copresented (with

Amy MacKinnon) at the Canadian Philosophical Association (virtually), June 2021, and the Philosophy of Science Association, Baltimore, November 2021.

Some of the people I have already mentioned have read and commented on parts of the manuscript, including Dylan Ludwig, Amy MacKinnon, and Josh Mugg. Sarah Robins also very kindly read a draft of Chapter 5 and lent me her insights into episodic memory; her writings on the subject have been a real source of inspiration for me. Brett Reynolds generously offered to read Chapter 6 and lend his expertise in linguistics, for which I am very grateful. Finally, two anonymous referees for Cambridge University Press provided a wealth of detailed and incisive comments. I have not managed to heed all their advice and answer all their questions, but I hope that I have gone some distance in that direction. At Cambridge, Hilary Gaskin was the model of efficiency and magnanimity, going out of her way to accommodate some of my demanding requests.

It is customary at this point to say that my greatest debt is to my family, but I'd rather not think of my relationship with them in transactional terms, because that would leave me feeling much worse off. Speaking of family, this book is dedicated to my father, Tarif Khalidi, a very philosophical historian. Our conversations and debates, on figures ranging from al-Farabi to Wittgenstein, and on topics as diverse as historiography and physiognomy, first sparked my interest in philosophy. He likes to say that he understands around 10 percent of what I write; I worry that that may be the only cogent part.

CHAPTER 1

Cognitive Kinds

O the mind, mind has mountains; cliffs of fall
Frightful, sheer, no-man-fathomed.
– Gerard Manley Hopkins, "No worst, there
is none. Pitched past pitch of grief."

Is there no way out of the mind?
– Sylvia Plath, "Apprehensions"

1.1 Introduction

What is the landscape of the mind? That is the question I aim to tackle in this book. This is an inquiry into the basic components of our mental makeup: What kinds of objects, states, capacities, events, processes, and other entities constitute the stuff of our mental life? As the book's title indicates, the scope is not the mental in general, but the cognitive realm in particular, which I take to be a subset of the mental or psychological realm. Although I will not attempt to demarcate the limits of the cognitive in detail, in what follows, I will attempt to say what characterizes cognitive phenomena, as opposed to other aspects of the mind and brain, later in this chapter (see Section 1.5, as well as Section 2.6). The inquiry is grounded partly in metaphysics and ontology, the philosophical investigation of the building blocks of the universe, and partly in the sciences, empirical research into the workings of the human mind. Since this is a book written by a philosopher, the latter is represented not in the form of original research but by means of distillations of recent empirical work on various mental items and an attempt to synthesize empirical work from different disciplines and subdisciplines. Integrating this empirical work with philosophical argumentation requires paying attention to the relevant literature in cognitive science, including psychology in its various branches (cognitive, developmental, social, and so on), linguistics, neuroscience, computer science, and related disciplines. Given the voluminous amount of work in these areas, it may seem presumptuous to take it all in, and I certainly do not aim to give a comprehensive account of the mental landscape. Instead, I plan to focus on a small number of paradigmatic cases. Of course, this type of integrative project also requires the careful philosophical work of making distinctions, clarifying concepts, and justifying claims with arguments. In this introductory chapter, I intend to lay out some of the philosophical groundwork that supports the argumentation that follows in later chapters. In particular, I plan to spell out the approach to ontology that I intend to take, and specifically the account of categories and kinds that I will adopt, which is naturalist, non-reductionist, and realist (as I will go on to explain).

Inevitably, when one investigates the mind these days, the brain is never far behind. Some would say that the entities constituting the mind are none other than those that comprise the brain, and that we are well on our way to discovering what these are. But despite the fact that there is indeed an intimate connection between psychological and neural entities, I will try to provide reasons for thinking that they are not one and the same and that the categories that pertain to one may not apply to the other. Though the focus will be on mental or psychological entities, their connections and relations to neural entities will often be invoked. To anticipate somewhat, one of the main themes of this book is that there is not always an identity – whether type or token – between psychological and neural constructs, and furthermore, that the validity of a psychological construct does not reside in its coincidence with a neural structure, mechanism, or process. In the neurosciences, there is currently considerable debate and a notable absence of consensus about how mental and neural entities relate to one another. Neuroscientists run the gamut, from those who advocate extreme reductionist positions that posit a "grandmother cell" (see Gross 2002) or a "Jennifer Aniston neuron" (Quian Quiroga, Reddy, Kreiman, et al. 2005) and locate cognitive functions (even particular concepts) in specific brain regions or populations of neurons, to those who preach anti-reductionism and excoriate "blobology," the alleged identification of areas of neural activation with particular psychological capacities, primarily based on regions identified by neuroimaging technology (Poldrack 2012). In subsequent chapters, I will try to provide reasons for thinking that though we are finding and will continue to find many significant correlations between brain structure and cognitive or psychological function, we should not expect a wholesale identification of one with the other. Indeed, I will argue that we will not always be able to identify psychological functions with neural activity, whether or not this activity is localized in specific neural structures. As I mentioned in the previous paragraph, the emphasis will be on *cognitive* ontology rather than psychological ontology more broadly. All the case studies to be discussed involve cognition in some way, as opposed to affective, perceptual, sensory, or experiential aspects of mentation. The aim is not to give an exhaustive catalogue of the contents of the mind (if that were even possible) but rather to focus on a range of significant examples of categories that involve cognition, examine the case for admitting each into our ontology, and draw some general conclusions about the kinds of entities that we should posit in cognitive science and on the grounds for doing so. After this first programmatic chapter, each of the rest of the book's chapters tackles one or a small number of candidates.

1.2 Naturalism about Kinds

In investigating mental objects, states, capacities, events, processes, and other entities, we are usually investigating types not tokens, that is, not unique particulars, but types or *kinds* of them. Specifically, we are interested in which of these types or kinds are real or "natural," or in standard philosophical parlance: *natural kinds*. Many contemporary discussions of natural kinds base their notion of kinds on the essentialist account first sketched out by Putnam (1975) and Kripke (1980). Instead, I will anchor the account of kinds that I will be deploying throughout this book in a nineteenth-century tradition that is more closely aligned with a naturalist philosophical outlook. According to the naturalist tradition that I will be tapping into, empirical science is our best guide to the kinds that exist in nature, rather than a priori considerations from metaphysics or philosophy of language. This attitude originates with the discussion of scientific classification that is prominently represented in the works of Whewell and Mill, and indeed in their mutual influence. Though Mill is often credited with initiating the discussion of natural kinds (or just plain "kinds," as he called them) in modern philosophy, even a casual reader of Mill's *A System of Logic* (1843/1882) cannot help but notice the considerable debt to Whewell's *Philosophy of the Inductive Sciences* (1840/1847). Despite significant differences in their overall philosophical positions, Whewell being a neo-Kantian rationalist and Mill a staunch empiricist, there is much that they agree on when it comes to kinds. Whewell and Mill both regard science as the guide to uncovering kinds in nature and think that scientific taxonomy aims at discovering kinds. Moreover, they are both concerned with the rational grounds for scientific classification and are keen to understand the differences between "natural" and "arbitrary" scientific classification schemes. They both see kinds as the basis for inductive inference and regard science's search for kinds as a quest to come up with categories that would lend themselves to empirical generalizations and natural laws. As Whewell writes: "The object of a scientific Classification is to enable us to enunciate scientific truths: we must therefore classify according to those resemblances of objects … which bring to light such truths" (1840/1847, 486). Whewell also thinks that classification must not be based on any resemblances whatsoever but on what he calls "natural affinity," which requires us to classify things on the basis of properties that generally cooccur with other properties (1840/1847, 542). Moreover, he repeatedly states that "the great rule of all classification" is that "the classification must serve to assert general propositions" (1840/1847, 495). Mill endorses this emphasis on "general propositions" or "general assertions" and goes on to say that "the very first principle of natural classification is that of forming the classes so that the objects composing each may have the greatest number of properties in common" (1843/1882, 879). Hence, for both Whewell and Mill, the aim of scientific classification is to group things together based on shared cooccurring properties, so that the categories that result enable us to make valid scientific generalizations.

While the naturalist tradition that originates with Whewell and Mill provides the main philosophical inspiration for the account of kinds that I will be operating with in this book, there is one respect in which I will part company with this older tradition. These philosophers are not very clear when it comes to the metaphysics of kinds. They seem to think that uniformities in nature are the basis for successful scientific generalization and inference, but they do not fully explicate the nature of these uniformities. Venn (1889/1907) criticizes Mill for distinguishing between two kinds of uniformity in nature: uniformities of sequence (which are causal) and uniformities of coexistence (which are brute). By contrast, Venn thinks that many of the uniformities of coexistence identified by Mill are actually causal in nature (though he does not think that *all* uniformities in nature are causally based). Still, he holds that uniformities are what enable us to use natural kinds in inductive inference in science. According to Venn (1889/1907, 94), uniformity "is the objective counterpart or foundation of inferribility …." Inductive inferences are based on uniformities and are therefore dependent on the existence of kinds in nature, which reflect these uniformities. Thus far, I agree with Whewell, Mill, and Venn. But by contrast with them, I will assume that uniformities in nature are due to regular and stable connections between causes and effects, and that these

1.2 *Naturalism about Kinds*

causal relations are the metaphysical bases of scientific induction and epistemic practices.[1] This assumption is also shared by many contemporary naturalist philosophers. It is prominent in Boyd's account of natural kinds and it is exemplified in what he calls the "accommodation thesis": "Kinds useful for induction or explanation must always 'cut the world at its joints' in this sense: successful induction and explanation always require that we accommodate our categories to the causal structure of the world" (1991, 139). Boyd also speaks of "the accommodation of inferential practices to relevant causal structures" (2000, 56).[2] This is also a central feature of Kornblith's (1993, 35) account of natural kinds: "It is precisely because the world has the causal structure required for the existence of natural kinds that inductive knowledge is even possible." This link between the epistemology of categories and the ontology of kinds is characteristic of a naturalist attitude to metaphysics, which holds that our metaphysical inquiries should be guided by our best epistemic practices as exemplified in the considered classification schemes of our best scientific theories. Among at least some contemporary naturalist philosophers, the causal structure of the world is the ontological basis for the successful epistemic practices of science.

Contemporary naturalist philosophers think that the causal uniformities in nature, even those discovered by the basic sciences, are rarely if ever ironclad or exceptionless, and this implies that the properties associated with natural kinds are loosely clustered rather than invariably associated with one another. Moreover, as I have already mentioned, the properties that cluster in kinds are not just sets of properties that happen to cling together, since they are associated as a result of causality. Accordingly, rather than view kinds as mere clusters of properties, I have proposed that they be conceived as "nodes in causal networks" (Khalidi 2013; 2018). According to this "simple causal theory" of natural kinds (cf. Craver 2009), certain properties or conjunctions of properties that are causally connected with others in systematic ways can be considered natural kinds. Sometimes we identify the kind with just one of the properties in a causal chain or network,

---

[1] There may be some uniformities in nature that are brute and not causally based, particularly at the most fundamental level. But I will assume that these are not at issue in a discussion of cognitive ontology. For further justification, see Khalidi (2013; 2018).

[2] Elsewhere in the same paper, Boyd emphasizes the ways in which natural kinds are "practice-dependent" and relative to human interests, and it is not easy to reconcile this attitude with his accommodation thesis. On the view that I favor, human interests serve only to select certain causal structures and processes to focus on, they do not somehow shape or modify them (except in cases in which humans are themselves part of the causal process – see Section 1.5).

but at other times we draw a wider circle among a number of them and consider that set of properties to be the kind. Either way, we are identifying properties that are causally conjoined to others, rather than mere clusters of properties. This causal account of kinds is somewhat less restrictive than that proposed by Boyd, who considers kinds to be property clusters that are held in homeostasis by causal mechanisms – though he sometimes relaxes these conditions and gestures toward something like a simple causal account. Thus, the simple causal account is distinct from a strict version of Boyd's account, which requires a specific causal mechanism to keep the cluster of properties in equilibrium (or homeostasis). I have questioned the strict version on two grounds. First, in many cases, there is nothing that can properly be called a causal mechanism that holds the properties together – they may instead be held together functionally or relationally, as we shall see in later chapters. Second, the properties involved are not always in a state of equilibrium – they may be repeatedly instantiated through the action of independent causes.[3] A simple causal theory of kinds can also be usefully distinguished from an essentialist one, at least on many versions of essentialism. Though essentialists also tend to think that natural kinds are discoverable by science, they usually place additional conditions on natural kinds, which I think are at odds with scientific taxonomy. There are four ways in which this account of kinds differs from many essentialist ones. First, the properties that are associated with each kind are causally linked, but they can consist in a loose cluster rather than a set of properties that are both necessary and sufficient for kind membership. Second, the causal properties may be functional or relational rather than intrinsic. Third, the properties involved do not have to be microstructural, as some essentialist philosophers tend to insist. Fourth, the simple causal theory does not claim that these properties are associated with the kind in question across possible worlds or with modal necessity, as essentialists usually hold.

Another significant point of agreement in the naturalist tradition that stems from Whewell, Mill, and Venn is that natural classification schemes and the kinds that they identify can be found across the sciences, including the human sciences. These philosophers tend to see considerable continuity from chemistry to mineralogy to biology to psychology and the social sciences, especially when it comes to the importance and feasibility of uncovering kinds. This attitude seems less prevalent among contemporary

---

[3] These claims are further justified in Khalidi (2013; 2018). I have also proposed that natural kinds can be represented by means of directed causal graphs. Although I have not worked out this proposal in detail, in such representations, natural kinds correspond to highly connected vertices in directed causal graphs.

1.2 Naturalism about Kinds

philosophers, at least some of whom think that it is a truism that natural kinds pertain to the natural sciences. Hence, it may appear oxymoronic to talk about natural kinds in the cognitive sciences. Given that the terminology of "natural kinds" is misleading, especially in the context of the human sciences, I will be talking mainly of "kinds" or "real kinds" instead of "natural kinds," especially given that the very existence of the expression "natural kind" seems to be a historical accident. As Hacking (1991) has pointed out, the terminology of "kinds," which Whewell and Mill used, gave way to "natural kinds" as a result of the writings of Venn. But Venn seems to have taken himself to be using Mill's expression, since he credits him with introducing the term – despite the fact that Mill apparently never used it. Venn (1889/1907, 84) writes: "he [i.e. Mill] introduced the technical term of 'natural kinds' to express such classes as these." It is unclear whether Venn simply misremembered Mill's terminology or whether he deliberately modified it. Either way, we are now saddled with an unfortunate expression, which is misleading on at least two counts. The first reason that the expression "natural kind" is deceptive is that it tends to set up a misguided contrast between the natural and *artificial*. In many scientific domains, there are strong candidates for kinds that have the "trail of the human serpent" over them and may reasonably be considered artificial (especially in the Anthropocene era). Whether we are dealing with synthetic chemicals, genetically engineered organisms, or artificially intelligent systems, scientists now study a range of entities that are the result of human intervention (if not wholesale invention), yet apparently no less real or objective than their supposedly "natural" counterparts. But the terminology of "natural kinds" would encourage us to dismiss the kinds to which these entities belong. The second reason the expression is misleading (which is more important for these purposes) is because the adjective "natural" suggests an affiliation with the natural rather than the social sciences, and it threatens to sideline categories that have a social or human dimension. When it comes to the cognitive sciences, which straddle the biological and psychological sciences, this is especially pernicious, since it tends to privilege the former over the latter, perhaps suggesting that neural kinds are more objective than psychological ones.

Here, it may be objected that the philosophical apparatus of real kinds may not be the right lens through which to view cognitive science. It may be thought that kinds are more at home in sciences like botany or mineralogy, where the paradigmatic individuals are well-defined concrete particulars (individual plants, mineral samples), with clear spatiotemporal boundaries. In cognitive science, though there are some fairly neat individuals such as human persons (and other creatures), which are often classified into kinds (e.g. *schizophrenic*, *bilingual*), the individuals can also be cognitive modules, cognitive capacities, mental states, mental processes, and other entities, so it may not be as useful to think of such entities as belonging to kinds. I would reply simply by stressing the indispensability of taxonomy to any scientific discipline or subdiscipline. Whenever we theorize about any domain, it is inevitable that we classify the items that populate that domain and that we do so in nonarbitrary ways. Classification, in turn, presupposes dividing a domain of entities into types or kinds. Moreover, as I will try to show, although some of the items classified in cognitive science are not best thought of as individuals, but states, capacities, events, processes, and so on, they are also divisible into kinds. Hence, there is no need to think of classification as pertaining exclusively to a domain in which concrete particulars with well-defined spatial boundaries are the main items of interest.

This brief sketch of a naturalist theory of kinds and its underlying metaphysics will have to suffice for now. More details will emerge as we survey a number of candidates for cognitive kinds in subsequent chapters.

1.3 Ontological Matters

In recent philosophical and scientific discussions of cognitive ontology, it is common to read that "ontology" is used differently by philosophers and others, namely psychologists, neuroscientists, and perhaps most prominently, computer scientists. I believe that this claim is not wholly justified. There are perhaps some differences in emphasis and nuance in the usage of these disciplines, but this is not a case of sheer polysemy. The main difference may be that computer scientists (in particular) are interested in how domains are taxonomized without great regard to how they *ought* to be taxonomized, and without a commitment to the domain's actually containing the entities that are posited by the taxonomic or classificatory system. Philosophers, on the other hand, tend to be interested in the *ought* and in the underlying structure of reality. As emphasized in Section 1.2, naturalist philosophers tend to think that our current, mature, scientific taxonomic systems are our best (defeasible) guides to that underlying structure. In other words, they derive an *ought* from an *is*.[4] This is warranted on the

---

[4] This is a stark and provocative way of putting it. For a more nuanced account of the relationship between scientific practice and philosophical theory, see Khalidi (2013), where I lean on the notion of "reflective equilibrium," first introduced by Goodman (1954/1979).

assumption that science aims at discerning that structure. In later chapters, we will encounter challenges to that assumption, on the grounds that some investigators are not just aiming to discern the causal structure of the world, since their inquiries are shaped by non-epistemic norms (especially in areas like psychiatry). I will put such concerns to the side for now, and will take them up in some subsequent chapters (but see also Section 1.5).

If we are naturalists, then talk of "ontology" is closely related to talk of "taxonomy" or "classification" – provided we think that science aims primarily at classifying entities in such a way as to discern the causal structure of the world, and is guided in doing so by epistemic goals. When viewed thus, there does not seem to be an equivocation or ambiguity in the use of the term "ontology" and related expressions. If we bear in mind that "ontology" should not be used as a synonym for "taxonomy" or "classification scheme," but rather to denote the metaphysical structure that is described by a taxonomic system or classificatory scheme, then some of the differences in usage can be cleared up. This caveat is also relevant to the use of terms like "kind" and "category." These two terms (and related ones) are often used interchangeably, by philosophers and cognitive scientists alike, but I propose to distinguish them, as follows. A *kind* should be understood to be an entity in the world, which can be conceived of as a collection of particulars or set of entities (nominalist reading), or an abstraction, such as a universal that is immanent in particulars (realist reading). Meanwhile, a *category* pertains to our conceptual, theoretical, or linguistic framework and practices; it is the concept of a kind.[5] In other words, a kind pertains to ontology whereas a category pertains to taxonomy. Here again, the two notions are closely related, since (on a naturalist understanding) the aim of scientific inquiry is to devise categories that correspond to all and only the kinds.

Once we distinguish ontology and taxonomy, along with kinds and categories, we should take care not to embrace a view that has been derided as the "third dogma of empiricism" (Davidson 1973). According to this dogma, we can somehow confront our kinds with our categories directly to determine whether they are in alignment, as we might compare a map of the landscape with the terrain itself for accuracy. The problem with this way of thinking is that we have no access to the "terrain" that is not mediated by our "map" (which is why the cartographic analogy is so misleading). We access the world via the categories of our taxonomies and hence, we cannot step outside of them to see how well they align with the

---

[5] To mark this distinction, I will generally italicize *kinds* and put CONCEPTS in small caps.

world itself. But that does not mean that we have no way of determining whether and how well our categories delineate the kinds. Since we devise these categories to describe the world, we can determine how well they enable us to generalize, explain, predict, and so on. Depending on their efficacy in fulfilling our epistemic goals, we infer that they have or have not latched on to the causal structure of the world, in line with the naturalist picture outlined in Section 1.2. This view has been articulated lucidly by Child (2001, 38), who writes: "in classifying things by reference to their causal powers, or their causally significant composition, we classify things in ways that reveal the way the world works." As long as human inquiry is able to achieve this goal, the specification of an ontology is not beyond our reach.

I argued in Section 1.2 that classification and the identification of kinds is based on identifying properties that are associated with other properties, and these properties are so identified because they are causally related. But what about those properties themselves? How do we identify the most basic properties in our ontology, and might we have settled on a different system of kinds if we had started with a different set of properties? This is an old philosophical conundrum and I cannot pretend to give a satisfactory answer to the question in the scope of this book (see e.g. Goodman 1954/1979; Lewis 1983). It is true that real kinds are grounded in shared properties and that these properties may be considered the unjustified posits upon which the whole theoretical edifice is built. If the properties to which we humans are attuned are just reflections of our parochial perceptual and cognitive abilities and do not reveal real features of the universe, then you might say that we have no reason to believe that the kinds that we identify expose the real joints in nature. In cognitive science, such properties might include basic behavioral ones involving motion, force, space, and time (e.g. eye movements, button presses, looking times, reaction times) or more abstract intentional ones (e.g. expressed preferences, discrimination between stimuli). But there may be a way of overcoming certain skeptical doubts about these baseline properties. For there is an indirect vindication of our choice of baseline properties in the identification of kinds that enable us to make generalizations, which in turn help us to explain and predict the entities in question. As I argued in the previous section, these epistemic desiderata are themselves causally based, so the choice of properties is ultimately upheld by our ability to use them to understand the causal structure of the world. Unless causality is itself an illusion, or a mere reflection of our inadequate and distorted perceptual and cognitive endowments, our choice of properties in cognitive science is at least indirectly supported by the fact that the kinds that they constitute allow us to discern the causal structure of the mind.[6]

Here, I should also emphasize that on the ontological framework I am adopting, there is no strict divide between properties and kinds. Although kinds are usefully thought of as loose collections or clusters of properties, some of the properties that we take to be associated with kinds can be further decomposed into or identified with more basic properties (perhaps also in the sense of loose clusters). Some properties are best thought of as complex ones that can be understood in terms of more basic properties (e.g. in physics, the property of being an electron decomposes into other properties, namely having a certain mass, charge, and spin). It is true that this way of thinking about properties and kinds purports to justify them with reference to each other, but circularity of this sort seems unavoidable in defending our most basic choice of parameters and dimensions. Moreover, it is evident from reflection on the history of science that the properties and kinds that we identify on the basis of preliminary observations and investigations are not always those that persist after subsequent inquiry. Therefore, it is not as though we are locked into our initial choice of properties no matter how misguided. This may not be enough to assuage skeptics or anti-realists, but I will have more to say about realism in the following section.

There is a further ontological question about properties and kinds, briefly alluded to earlier, that I would like to bracket as far as possible in the context of this book. That has to do with the traditional metaphysical debate between nominalism and realism, which pertains to the underlying metaphysical reality of properties and kinds. Are they universals (metaphysical realism) or collections of individuals (nominalism)? Although I am committed to the existence of properties and kinds, I am not wedded to a particular metaphysical understanding of them. They may be best identified with universals, whether immanent or transcendent, or they might be better understood as sets or collections of particulars. Moreover, if they are identified with universals, it may be that kinds are universals in their own right in addition to properties, or they might just be concatenations of property universals. We need not resolve these questions for the

---

[6] What if causality itself is not objective? Causal connections are the holy grail in science, as witnessed by the fact that scientists are generally intent on disentangling causation from correlation. I am therefore assuming that causal connections are something like the skeletal frame of reality. Moreover, if there is no *unique* causal relation, but a plurality of causal relations (see e.g. Godfrey-Smith 2010), then there would be no single metaphysical basis for real kinds but a plurality. I will set this possibility aside for the purposes of this book.

purposes of this work, since I doubt that it matters much which position we adopt when it comes to cognitive ontology, just as long as we acknowledge the existence of properties and kinds.

Finally, on the topic of general ontology, I will not try to settle a question as to the identity of the broadest ontological categories. I have referred already to such entities as individuals (or objects), states, processes, events, capacities, and mechanisms, and have indicated the possibility of the existence of others. Such posits are sometimes considered the basic "categories" of ontology. Some philosophers posit a more austere basic ontology. For example, Lowe (2006) posits a "four-category ontology," in which the categories are individual substances (objects), substantial universals (kinds), property or relation instances (modes), and non-substantial universals (properties or relations). Other philosophers think there are other ontological categories that are not equivalent to one of these four, such as processes or events. I will not try to settle these metaphysical disputes, since the focus here is on the applied question of cognitive ontology. In the cognitive sciences, many entities are best conceived of as *individuals* (e.g. person), others as *states* (e.g. belief, pain), others as *processes* (e.g. learning, episodic memory retrieval), and yet others as *capacities* (e.g. language, semantic memory, mindreading). For other cognitive entities, there may be some uncertainty as to which broader ontological category they fit into. For example, we might well wonder whether *fear* is best understood as a *state* or a *process*. In some cases, there may be closely related entities that fit into more than one ontological category. For example, *episodic memory* is sometimes thought of as a *state*, *process*, or *capacity*, or indeed, all of the above – in which case the term "episodic memory" would be equivocal (for discussion, see Section 5.2). These questions will be explicitly addressed when it comes to some of the cognitive entities to be discussed in later chapters. In this book, it will emerge that the broader ontological categories that seem indispensable to cognition are those already mentioned: *individual*, *state*, *process*, and *capacity*. The cognitive kinds to be discussed in subsequent chapters all seem to belong to one of these four basic ontological categories. But I will not try to come up with an exhaustive list of broader categories comprising all cognitive entities. I will also not take a stand on whether some of these broader categories should be understood in terms of others, or whether some of them are more fundamental than others. For example, some metaphysicians would advocate reducing all processes to sequences of temporally ordered events, while others would consider processes to be fundamental and would argue for understanding events in terms of processes, while yet others would favor explicating all other metaphysical categories in terms of processes (thus advocating what has been called a "process ontology"). Since this is not a work in general ontology, I will not try to resolve these issues or take a definite stand one way or the other. If there are theoretical and empirical reasons for considering a certain cognitive entity to be a process rather than a state, I will try to indicate them; if some entities can be understood in more than one way, I will try to justify that claim. Throughout, I will just assume that cognitive entities may fit into a number of different broader categories and will give reasons for considering the particular entities to be discussed to fit into one or more of these overarching categories (e.g. process or capacity), without attempting to give an exhaustive list of such categories or trying to determine which of them (if any) are fundamental.

1.4 Reductionism

There is a very strongly entrenched picture of the domains investigated by science that portrays them as constituting something like a strict hierarchy, from that of elementary particles at the bottom of the hierarchy to social entities like nation-states and economic markets at the top. The idea is that the universe is arranged in a number of layers or levels ranging from the smallest and most fundamental to the largest and most complex. The picture is sometimes referred to as the "layer-cake view," though it is perhaps better described as the "inverted pyramid view," to convey the idea that the more fundamental domains are more austere in terms of the number of entities they posit and the laws or regularities that govern them, while the less fundamental ones are more prolific, not to say profligate. This picture is sometimes traced back to a classic paper by Oppenheim and Putnam (1958), but its prevalence and appeal would seem to go beyond the influence of any one particular source. Although it is undeniably the case that the universe consists both of micro- and macro-entities and that the latter are composed of the former, this picture is misleading as an account of the most salient divisions in nature and the manner in which the different parts of the universe fit together.

What is wrong with the view that reality is arranged hierarchically in a series of domains, the entities of each domain being composed of those in the domain below it, and with the accompanying idea that the theories that purport to describe and explain each domain are reducible to those that describe the domain below it? There are at least three respects in which the view is highly misleading, or at least prejudicial to certain philosophical positions regarding the relationship between the sciences, in particular those that study the mind and brain. First, it implies that the crucial dimension distinguishing scientific domains is that of size or spatiotemporal dimension. Second, the layer-cake picture gives the impression that there is a neat compositional hierarchy in which the domain studied by each scientific discipline or subdiscipline decomposes into and depends entirely upon the one immediately prior to it in the hierarchy. Third, and most importantly, the layer-cake picture encourages a reductionist account according to which the key to understanding each scientific domain lies in the domain that precedes it. It promotes the view that to understand what is really going on in chemistry, one must turn to physics, and to understand what is really happening in biology, one must revert to chemistry, and so on.[7] I will now put forward considerations against each of these claims in turn.

On the layer-cake view, the layers of the cake are often thought to correspond to spatial or spatiotemporal dimensions, with lower levels consisting of micro-entities that operate over very small timescales, while higher levels are composed of entities at increasingly larger spatiotemporal scales. However, the causal systems described by different scientific disciplines or subdisciplines span different spatiotemporal scales. The theories of fundamental physics describe elementary particles as well as galaxies (and many systems in between). To use an example from the cognitive sciences, if we are investigating the psychopharmacological effects of lithium compounds on stabilizing the moods and altering the behaviors of people suffering from mental illness, we may need to refer to the role of lithium ions in increasing the release of neurotransmitters at the molecular scale while also describing the behavior of individuals in macroscopic terms. The causally relevant processes in nature often occur in transverse sections that cut across spatiotemporal dimensions.

A second problem with the layer-cake view is that the domains of scientific theories or disciplines cannot be considered to be arranged in a strict mereological sequence or dependence hierarchy, since there is often only partial overlap between one domain and another and they cannot be conceived of as related compositionally (see Figure 1.1 for a crude attempt to illustrate the difference between the layer-cake view and an alternative picture). The inquiry just described, which relates to the

---

[7] In fact, Oppenheim and Putnam (1958, 15) explicitly support their reductionist picture by referring to the relation between psychology and neuroscience: "it has proved possible to advance more or less hypothetical explanations on the cellular level for such phenomena as association, memory, motivation, emotional disturbance, and some of the phenomena connected with learning, intelligence, and perception."

1.4 Reductionism

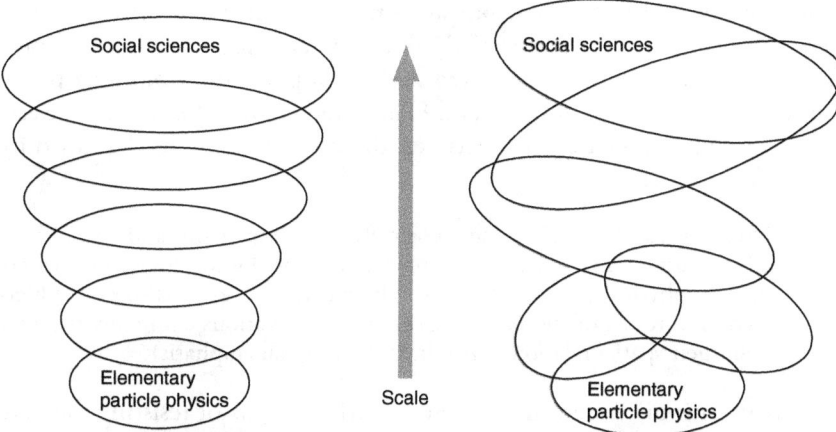

Figure 1.1. Alternative to the "layer-cake" view: The standard "layer-cake view" of scientific domains (left) can be contrasted with a picture (right) according to which scientific domains both crosscut one another and cut across spatiotemporal scales.

psychopharmacological effects of lithium, may be at cross purposes to another neuroscientific inquiry dealing with color processing in the visual system, for example. Both inquiries can be conceived of as investigating causal systems, but the systems involved are not wholly contained within one another. Even though in this case, both systems seem to depend entirely upon the neurobiology of the brain, I will argue later in this section that this impression is mistaken.[8] Rather than a series of ordered hierarchical levels, each of which depends on the next smaller level, an alternative picture is one in which the universe consists in a set of relatively "closed systems," each of which is causally integrated and somewhat causally cordoned off from others. The terminology of closed systems is derived from the physical sciences, but it should not be taken literally, since causal systems are not hermetically sealed. In the following chapters, I will loosely apply the expression to causally integrated systems in which the causal inputs are largely known and other external influences can be safely ignored or bracketed. Sometimes these coincide with scientific

---

[8] This does not just pertain to the relationship between the cognitive and neural domains. As noted by Rabin (2018), it is not clear how to order disciplines like geology and psychology in terms of dependence relations: both seem to depend on chemistry (and I would add, physics), but neither depends on the other. Similarly, Epstein (2009; 2015) argues that social facts do not depend entirely on facts about individual persons, or even those facts plus facts about their local environments and practices.

disciplines or subdisciplines, but at other times they comprise scientific research programs. This way of thinking of scientific domains is closely allied with a conception elaborated by Weiskopf (2017b), building partly on work by Simon (1969/1996) and Haugeland (1998). Weiskopf contrasts two views of the relationship between the complex systems described by scientific models:

> One way sees hierarchies mereologically, in terms of size and spatial containment relations, so that a system is decomposed into subsystems that are literally physical parts of it. … An alternative, however, is to define hierarchies in terms of the *interactional strength* of various components rather than their spatial relations. (2017b, 11–12; original emphasis)

My main (minor) dissent from these remarks consists of resisting the urge to posit a hierarchy at all. If interactional causal strength (and relative causal isolation) is what delimits a domain described by a scientific theory or model, there should be no expectation that scientific domains will constitute an ordered series or sequence, and hence there is no question of a hierarchical arrangement. This anti-hierarchical position also agrees with the attitude expressed by Woodward (2017, 40 n.4), who writes that he takes a "very deflationary" understanding of "levels," according to which "levels talk is just a way of expressing claims about explanatory or causal relevance and irrelevance," which "does not carry any suggestion that reality as a whole can be divided into 'layers' [or] levels on the basis of size or compositional relations"[9] Ideally, "levels" discourse should be entirely replaced with talk of causal systems or domains, but this discourse is entrenched in both philosophy and science and not easily expunged. I will use it occasionally in subsequent chapters, bearing in mind the caveats just mentioned, but I will usually replace it with talk of domains or causal systems.

The third problem with the layer-cake view is the association with reductionism. Classically, reductionism was understood in terms of a complete translation of the theoretical terms of one theory into the theoretical terms of another (Nagel 1961). But one need not hold reductionists to such a high standard. My objection is not just aimed at this classical view

---

[9] These views are also reminiscent of Wimsatt's prescient account of "levels" (first published in 1994). He writes that levels *"are constituted by families of entities usually of comparable size and dynamical properties, which characteristically interact primarily with one another, and which, taken together, give an apparent rough closure over a range of phenomena and regularities"* (2007, 204; original emphasis). Wimsatt conceives of levels largely along compositional lines, though he also argues for the existence of more complex and non-compositional "causal thickets," which I will mention in Section 1.5.

but to weaker notions of reductionism as well, which hold merely that lower-level explanations are more fundamental than higher-level ones, or that one has not fully explained or understood a particular domain unless one has explained it in terms of a lower-level domain. The criticisms I have made of the layer-cake view already suggest that we cannot expect that each domain will be wholly explainable in terms of a domain that is "lower" in the hierarchy, since the whole notion of a hierarchy is problematic. If one thinks of the domains of different theories or disciplines as being partially overlapping, depending on the causal integration and isolation of the entities under investigation, then that threatens not just a Nagelian reduction, but any form of reductionism that posits that each domain is best explained by the next lowest domain in the hierarchy. It might be objected that as long as there is one all-encompassing lowest domain, currently thought to be the domain of elementary particles, it should in principle be possible to reduce each domain directly to that fundamental base. But the reduction of special-science domains like cognitive science directly to fundamental physics is not a serious prospect.

In the rest of this section, I will try to bolster the case for denying even a weak version of reduction when it comes specifically to cognitive science. The usual way of resisting reductionism leans on the claims of multiple realizability and multiple realization. These topics have been extensively addressed in recent philosophical debates and I will not try to do justice to those debates. In brief, multiple realizability claims that cognitive and psychological categories are individuated differently from neural and biological ones. Since the former are typed in terms of their functions or causal roles but the latter are typed in terms of their structures or mechanistic properties, and since the same functions can be performed by mechanisms with very different structures, they cannot be type identical. The argument proceeds at a very abstract level based on what we know generally about functions and structures. Moreover, it concludes only that the kinds of psychology are multiply realiz*able* relative to neurobiology, not that they are actually multiply realiz*ed*. For all this argument claims, in our actual universe, the same psychological and cognitive functions are always achieved by the same neurobiological structures. The argument from multiple realization is subtly different. It relies on direct empirical evidence and bases its conclusion on purported cases in which the same cognitive state or process is subserved by different neural mechanisms. If one can point to actual cases in which the very same type of psychological function is performed (in different individuals, species, or systems, or indeed in the same individual on different occasions) by genuinely different types of neurobiological (or other) structures, then that blocks a type–type reduction. In a way, multiple realizability is the stronger argument since it is supposed to apply regardless of the empirical evidence and cannot be refuted by showing that alleged cases of multiple realization do not in fact support the desired conclusion. But in another way, multiple realization is more powerful, since any genuine finding of different structures realizing the same function would seem to clinch the case and render the abstract argument superfluous. Usually, the two go together, since multiple realizability is a plausible explanation for multiple realization, and multiple realization provides corroborating evidence for multiple realizability. Though these theses have had convincing advocates, they have also had vocal opponents, who claim that neither thesis has been demonstrated. Both theses have been exhaustively debated in the philosophical literature and I have nothing to add to those debates except to endorse the arguments that have been made in favor of multiple realization and multiple realizability.[10] I will not try to reiterate them or advance them here because I think that one can go even further in an anti-reductionist direction, as I will try to explain. In the rest of this section, I will try to provide an argument as to why the cognitive domain is not likely to neatly decompose into the neural domain.

Recent advances in neuroimaging have opened the door for scientists to investigate the workings of the brain in real time, while experimental subjects are engaged in various cognitive tasks. At one point, this development held out the hope for assigning specific cognitive functions to brain regions in a fairly direct and straightforward way. But despite the fact that the past few decades have revealed a great deal about the workings of the brain and how it serves to implement the functions of the mind, there is mounting evidence that the links between neural structure and cognitive function are rather more circuitous and indirect than has often been assumed. At least as things currently stand, the vaunted "structure-to-function" mapping is

---

[10] For a comprehensive defense of multiple realization using examples from vision science, see Aizawa (2017), where he argues that multiple realization results in a kind of autonomy for higher-level taxonomies, in the sense that there need not be an isomorphism between higher- and lower-level taxonomies. He also argues that this is compatible with the idea that there are important interactions between higher- and lower-level sciences, which constrain the development of their respective taxonomies. Multiple realization and realizability have also been vigorously disputed by some philosophers, for example, Polger (2002; 2009) and Shapiro (2004), but I will not try to address their arguments directly. Sober (1999) is also sometimes interpreted as an argument against multiple realizability, but his position seems more nuanced. He writes: "The reductionist claim that lower-level explanations are *always* better and the anti-reductionist claim that they are *always* worse are both mistaken" (Sober 1999, 560; original emphasis). But I do not take the anti-reductionist import of multiple realizability to be that lower-level explanations are always worse, just that they sometimes capture causal patterns that are not captured at the lower levels.

far from a one-to-one correspondence. To pick one example out of a hat, a circumscribed and well-defined part of the brain such as the hippocampus, which is a bilateral structure in the medial temporal lobe (MTL), has long been thought to serve the functions of episodic memory and spatial navigation, and historically, this was based primarily on evidence from lesion patients and animal models. But neuroimaging evidence suggests a far more expansive functional repertoire. According to recent work, there is evidence that the hippocampus plays a role in aspects of perception, attention, working memory, language, and semantic processing, "all of which were originally believed to be outside the domain of hippocampal and MTL function" (Kwan, Craver, Green, et al. 2015; see also references therein). It is tempting to think that there may be some common function or set of functions that is common to all these cognitive capacities, which is the unique cognitive function performed by this particular neural structure. But in this and many other such cases, there does not appear to be a common cognitive function, at least not one that can be specified in terms of our existing cognitive categories.[11] Furthermore, many cognitive functions or capacities are themselves subserved by numerous other brain regions and structures. Hence, what is emerging from these noninvasive neuroimaging technologies is evidence for a many-to-many relationship between neural structure and cognitive function. One of the most promising attempts to understand and explain this relationship is the theoretical framework of "neural reuse" (Anderson 2010; 2014). Neural reuse says that brain structures are "used for diverse purposes in various task domains" (Anderson 2014, 9). This means that when a population of neurons is involved in different cognitive tasks they do not always perform the same function. The cognitive function of a neural population is not an intrinsic property of that particular group of cells but depends on the patterns of interaction between those biological entities and others, in a way that is not yet fully understood in contemporary neuroscience. Neural reuse is therefore a theoretical proposal that urges neuroscientists to take into account not just "neural real estate" but the interrelations between sets of neurons in attempting to understand the relationship between brain and cognition.

---

[11] Some neuroscientists have advocated a thoroughgoing revision of our cognitive ontology in an attempt to locate the cognitive functions served by different neural regions or networks (e.g. Poldrack & Yarkoni 2016), but this effort has yet to yield convincing results. Anderson (2014, 128) sometimes argues that neural structures or regions may have lower-level functional profiles, which he labels "neuroscientifically relevant psychological factors" (NRPs), though I take it that these are not full-fledged psychological or cognitive functions.

If the support for neural reuse were limited to neuroimaging evidence based on fMRI technology, there would be an obvious way to resist the conclusion. It would be open to a skeptic to say that the reason that we are not always able to find a direct link between neural structure and cognitive function is that our current technologies are as yet unable to zoom in precisely enough or that we have not been careful enough to distinguish different areas within a supposedly multifunctional region. In the face of evidence of multifunctionality, proponents of "localization" sometimes adopt a "divide and conquer" strategy, which posits that different functions are being performed by different subregions of the same region (e.g. Saxe, Brett, & Kanwisher 2006; see also McCaffrey 2015b). To return to the example of the hippocampus, it may be thought that the diverse cognitive functions associated with this brain structure may on closer inspection turn out to be subserved by different parts or areas of the hippocampus. Even though the average volume of each half of the hippocampus on the right and left sides of the brain does not exceed a few cubic centimeters, that tiny volume includes tens of millions of neurons, and there would seem to be ample opportunity for different populations of neurons to specialize in different tasks. But Anderson (2014, 30–34) and other researchers have presented various considerations to support the conclusion that reuse persists no matter how fine-grained our investigative techniques. For example, he cites evidence that single neurons in the roundworm *C. elegans* sometimes perform both motor and sensory functions and that other individual neurons even participate "in generating completely opposite behavior as a result of alteration of the neuron's sensitivity, physical connections, and functional connectivity by various chemicals and genes (effects known collectively as neuromodulation)" (Anderson 2014, 32). This and other types of evidence would undermine the "divide-and-conquer" strategy favored by deniers of neural reuse.

Multiple realization and neural reuse can be seen to be mirror images of one another, as a quick comparison will show. According to multiple realization, there is a one-to-many relationship between mind and brain. One has to be careful about stating this claim, since it is not always clearly distinguished from a substantively different claim. Proponents of multiple realization are not merely saying that some cognitive capacities are subserved by a number of different brain regions working in concert. If that were all there was to it, multiple realization would be no threat to reduction. Reductionists could simply reply by saying that the unit of interest is not the neural region but the set of regions or neural network, and that when these units are properly individuated, they might well be put in a one-to-one correspondence with cognitive functions. Rather, multiple realization is saying that different neural structures can perform the same cognitive function, as it were, in parallel.[12] This is not a matter of different neural structures teaming up within the same system, but the same type of region or network or other configuration doing so in different systems (e.g. species, individuals, or even individuals-in-context), despite the fact that they are not type identical. Multiple realization blocks reductionism because it says that the relationship between mind and brain is one-to-many. Neural reuse, on the other hand, declares the relationship to be many-to-one, since it holds that there may be multiple cognitive functions that are subserved by a single neural region or structure. In some ways, this claim is more puzzling than that of multiple realization, since it seems to be denying something like a basic principle of metaphysical supervenience, namely that there can be no mental difference without neural difference. But as can be gleaned from the brief sketch of neural reuse above, the claim is that the functions of neural structures do not attach to them intrinsically, but are partly a result of extrinsic factors such as their anatomical and functional connectivity, as well as modulation by genes and chemicals. Hence, neural reuse is saying that cognitive function does not pertain to populations of neurons but arises out of complex interactions in the brain. In principle, this does not block reduction outright since if neuroscientists are able to identify all the relevant interactions and modulations, it may be possible to find the neural correlates of cognitive functions, even though these are not simply identical to regions or networks. Some of the neural factors may not be such as to be readily identifiable using existing neuroimaging techniques, but they are presumably identifiable in other ways. However, I will now put forward what I take to be a more principled obstacle to a reduction of the cognitive domain to the neural domain.

If held jointly, the claims of multiple realization and neural reuse issue in the assertion that the relationship between cognitive function and

---

[12] For instance, Price and Friston (2005, 262) do not sufficiently distinguish the two senses of a one-to-many mapping in the following passage:

> Functional neuroimaging data preclude a one-to-one mapping in two ways. First, attempts to manipulate a 'single' cognitive process (e.g. semantics) often elicit a distributed pattern of activation over many areas (i.e. a one-to-many mapping *from function to structure*). Second, the same brain region, or set of regions, may be activated by tasks with different cognitive processes (i.e. a many-to-one mapping). In short, there is a many-to-many mapping between cognitive functions and anatomical regions, with a range of cognitive processes emerging from different patterns of activation among a limited number of brain regions.

Here, the first case they cite, of a one-to-many mapping, is not a case of multiple realization as it is usually understood.

neural structure is many-to-many. It is not uncommon for neuroscientists to explicitly endorse a many-to-many mapping in recent years (e.g. Price & Friston 2005; Poldrack & Yarkoni 2016), especially as neuroimaging evidence has piled up indicating a convoluted relationship between structure and function. While I think there is strong evidence to support both claims, multiple realization and neural reuse, I will try to make the case against mind–brain reduction a little stronger. One of the broad aims of this book, to be supported by looking more closely at specific case studies in subsequent chapters, is to argue for a many-to-one relationship between mind and brain, but not just for the reasons provided by the advocates of neural reuse. In addition to the fact that specific neuronal populations appear to be implicated in diverse cognitive functions due to neural connectivity, neuromodulation, and other relational factors that are internal to the brain, I will argue that another reason that neural regions do not always subserve the same cognitive functions is that these functions are individuated in part with reference to the surrounding context of the thinker and the thinker's history. Since cognitive functions are often individuated externalistically or anti-individualistically, while neural structures, mechanisms, and processes are usually not so individuated, I will argue that this blocks the possibility of a correspondence between cognitive categories and neural ones, even taking into account the subtleties of brain chemistry and electrophysiology (see also Khalidi 2017; 2020). The claim is not that neuroscience never has occasion to individuate its constructs relationally or etiologically, but rather, the point is that even when it does engage in relational or etiological individuation, the neuroscientifically salient relata or causal histories do not always coincide with those relevant to cognition.[13] In subsequent chapters, I will try to show that some of the principal categories that we use to explain and predict cognitive phenomena are individuated with reference to contextual and etiological factors that are not generally invoked in understanding the workings of the brain, and this leads to a mismatch between the taxonomic systems of, say, cognitive psychology and neuroscience. The claim is not just a causal

---

[13] For a defense of etiological individuation in neuroscience, see Garson (2011), specifically when it comes to ascribing functions based on evolutionarily selected effects rather than synchronic causal roles. However, he also acknowledges that non-etiological individuation "appears to be more consistent with neuroscientific practice, which is more preoccupied with structural and functional decompositions of complex abilities than with speculation about evolutionary histories" (Garson 2011, 549). Amundson and Lauder (1994) put forward a general defense of ascribing synchronic causal role functions in various areas of biology, such as comparative anatomy, and at least some of their arguments would seem to apply to neuroscience.

1.4 Reductionism

but an individuative one. It is not merely that many cognitive states are both effects and causes of environmental factors (since many neural states are too). Rather, cognitive entities are often individuated in part on the basis of such factors, whereas neural entities are usually not so individuated, and this is what leads to a many-to-many mapping between neural and cognitive taxonomic categories. This results in crosscutting systems of categories, ruling out even a local or limited reduction.[14]

In one sense, it should not be surprising that there can be a many-to-many mapping among neural structures and cognitive functions. That is what we find in other scientific domains when we are investigating structures and functions, particularly in the biological sciences. It is widely accepted that different biological structures can perform the same function, as when arthropod eyes and vertebrate eyes perform the function of sight and "display similar psychophysical phenomena" (Weiskopf 2011, 236; cf. Stinson 2016). It is also relatively uncontroversial that the same biological structure can perform different functions, as when bird feathers are exapted from performing the function of thermoregulation to serving the function of flight (cf. Ereshefsky 2012). In general, then, the structure-to-function mapping in biology can be expected to be many-to-many, rather than one-to-one, or even one-to-many. This also holds for brain structures and cognitive functions in particular. When it comes to multiple structures performing the same function, advocates of multiple realization have brought forth various examples and considerations to support the conclusion that the same cognitive function can be performed by different neural structures in different species, in different individuals, and even in the same individual at different times (see e.g. Aizawa & Gillett 2009). As for multiple functions being performed by the same structure (i.e. multifunctionality), the arguments for neural reuse emphasize the fact that it is not just structures that account for functions, but also the relations between structures and the ways in which those structures are modified chemically and by other means. To further support the claim of multifunctionality, I would point to the fact that a thinker's environment and causal history can enter into the individuation or identification of the cognitive function being performed. In some cases, the same neural

---

[14] Although there has been discussion of the tension between the externalist and internalist individuation of mental states at least since the 1970s, many philosophers have assumed that individuation in a science of psychology, as in neuroscience, is always internalist or individualist (Stich 1985; Fodor 1987), with few exceptions (notably, Burge 1986). By contrast, I will go on to argue that the specific type of externalism that I will be arguing for is widely attested in the study of cognition and that it creates obstacles for a general type reduction among the kinds of psychology and neuroscience.

structures can be implicated in different cognitive functions because those functions are individuated with reference to environmental and etiological factors. This means that cognitive kinds are often typed differently than neural kinds. Moreover, this applies not just to neural *structures* but to the *totality of neurophysiological facts*, since these are not generally typed with reference to a thinkers' environment and causal history. Although some neural objects, states, processes, mechanisms, and other entities are sometimes individuated environmentally and etiologically, much of neuroscience investigates the brain in relative isolation from the environment and from its evolutionary history. Even when neuroscience is interested in identifying functional properties, the environmental and etiological factors that are relevant to neuroscientific taxonomy are not always the ones that are pertinent to cognition.[15] Hence, multi-functionality (or the one-to-many mapping between neural and cognitive categories) is supported not just by neural reuse but by the individuation of cognitive kinds with reference to certain distinctive environmental and etiological factors. This type of multi-functionality goes beyond considerations drawn from neural reuse and suggests that the one-to-many mapping applies to neural categories in general, not just to structural categories, such as brain regions or neural networks (e.g. *hippocampus, default mode network*). Moreover, if cognitive facts do not just supervene on neural facts, this would rule out not just a type reduction, but an extremely weak dependence of cognitive facts on neural facts. It would also mean that neural generalizations and explanations cannot be expected to wholly account for cognitive entities. This is not a limitation of neuroscience, just a result of the fact that different disciplines and subdisciplines investigate different causal systems, as indicated earlier.

1.5 Realism

Philosophers sometimes make a distinction between classification on the basis of nature's own divisions and classification on the basis of human interest. But there would seem to be something a little misguided about

---

[15] Craver (2013) defends the importance and centrality of a non-etiological conception of function to neuroscience (and the physiological sciences more generally). For example, he points out that a type of molecule can be described as a neurotransmitter, regardless of "the developmental or evolutionary origins of the molecule in question" (Craver 2013, 137). Garson (2019) disputes that there is a useful non-etiological conception of function in the biological sciences. For my purposes, however, what matters is that the environmental and etiological factors relevant to neural functions are often different from those relevant to cognitive functions.

this alleged contrast, at least if our interest lies in making valid distinctions. If epistemic conditions are good and if we are not simply cognitively incapable of understanding some aspects of the world, then our considered theories can be expected eventually to converge on the causal structure of the world. These theories comprise the classification schemes that purport to reveal the real kinds that exist in the world. I have argued elsewhere that our epistemic interests are geared toward discovering causal structures and that as long as inquiry is guided by these interests (as opposed to moral, political, aesthetic, or other interests), then there is no conflict between classifications that serve human interests and those that aim to uncover the divisions in nature (Khalidi 2013). This should apply to the human mind no less than to any other corner of the universe. This is how the naturalist picture sketched out in previous sections meshes with realism about scientific categorization and the search for real kinds. In the previous section, I defended an anti-reductionist position regarding cognitive entities, and in this section, I will try to show that this position is compatible with realism about cognitive kinds. I will do so by attempting to respond to two challenges that might be thought to confront an anti-reductionist realist position about cognitive kinds: mind-dependence and self-reflexivity. But before doing so, I will briefly address a question that pertains to realism not just in the cognitive domain but in other special-science domains.

The categories in the fundamental sciences, such as elementary particle physics or elemental chemistry, seem quite distinct: their individuation conditions can be clearly delineated and they can be easily distinguished from one another. These categories may not be as crisp as many philosophers have traditionally assumed, yet they appear to have a "naturalness" that categories in many other sciences appear to lack. Meanwhile, categories in the biological, psychological, and social sciences, do not seem to be as clearly delineated and they do not appear to be as distinct from each other. Moreover, I think it is safe to say that this is not just a feature of the categories in these respective sciences but of the kinds themselves. In other words, it is not an artifact of our taxonomic practices in these sciences but a property of the underlying kinds. How can we account for this difference? One way to think about it is to relate it to the nature of causal relations in the different domains. Though causal relations appear to be deterministic at the macro-level, there are many more intervening causes in the causal systems that occur in the domains of the special sciences. This difference has largely to do with the asymmetrical relationship between the micro-level and macro-level, since domains that tend to depend constitutively, causally, and otherwise, on other domains tend to be more complex and involve many competing causal processes. I mentioned in Section 1.4 that scientists try to isolate "closed systems" in their areas of inquiry, but this is easier done in a test tube than in the psychology lab. That is not just because one can spatially isolate the causal process of a chemical reaction more fully than that of a participant performing a cognitive task, but also because there are fewer other causal processes that might interfere with it. When it comes to the chemical reaction, the temperature in the room may do so, as may the level of humidity, and perhaps the lighting, among others, but the potential intervening factors are far fewer than those that might intrude on the psychology experiment. When it comes to the latter, in addition to the preceding factors, we might also mention the facial expression of the experimenter, her tone of voice, and music from the hallway, not to mention what the participant dreamt the night before, as well as his heart rate, psychiatric diagnosis, cultural background, gender, and level of education, among many others. On the conception of kinds that I have spelled out, this difference in the nature of causal processes in different domains leads to a difference in the nature of kinds. Such domains are dubbed "causal thickets" by Wimsatt (2007, 237), and he characterizes them as follows: "With increases in the complexity of objects, and in their number and variety of degrees of freedom, they can interact with one another in more varied and complex ways ..."[16] This means both that the causal connections between properties is more messy (due to possible interventions by other causes) and that there are more potential candidates for kinds depending on which combinations of properties are chosen for singling out and which causal processes we investigate. (For instance, should we make a rough psychological generalization over all humans, or a narrower generalization over a smaller reference class of humans with a certain cultural background? Should we regard, say, *male humans* or *Western humans* as subordinate kinds for certain psychological purposes?) It also means that there are often a number of different kinds in the same vicinity and we can choose which ones to focus on in our scientific theorizing. That does not mean that the boundaries between kinds are arbitrary, but it does suggest that we are sometimes at liberty to shift them somewhat while continuing to capture real kinds, that is, clusters of properties causally linked to other properties. Does this provide some room for bringing non-epistemic interests to bear on the demarcation of kinds? I would argue that these taxonomic choices will generally serve different *epistemic*

---

[16] Wimsatt (2007, 238) writes explicitly that "[t]he neurophysiological, psychological, and social realms are mostly [causal] thickets ..."

1.5 Realism

interests and that the demarcation of taxonomic boundaries is not driven by non-epistemic interests. That is because different causal processes are captured by different demarcations, which will deliver different explanations.[17] To provide further support for this claim, I will try to illustrate it in examining some of the cognitive kinds discussed in later chapters.

Here, something more needs to be said about the causal system that coincides with the cognitive domain. As I conceive it, the domain of cognition lies broadly within what Marr (1982) famously identified as the computational level, as opposed to the algorithmic and implementational levels. On Marr's view, the computational level of explanation "specifies what and why," whereas the algorithmic level "specifies *how*" (1982, 23; original emphasis). Meanwhile, the implementational level investigates "the particular mechanisms and structures ... in our heads" (1982, 19). Marr was explicit in holding that the computational level can be multiply realized in the algorithmic, as can the algorithmic in the implementational. As he put it: "the same algorithm may be implemented in quite different technologies" (1982, 24). A theoretical explanation at the computational level characterizes a cognitive system in informational terms, taking one kind of information as input and transforming it into another. Computational explanations also demonstrate why that informational mapping is appropriate and adequate for solving the problem or performing the task at hand (Marr 1982, 24). To use an example that will be briefly encountered in Chapter 4, vervet monkeys have an alarm call system to alert conspecifics to the presence of three different types of predators: leopards, eagles, and snakes (Cheney & Seyfarth 1990). The vervets give a different response to each type of predator and conspecifics react by taking the appropriate evasive action. A computational or cognitive theory of the system would describe the different inputs that the vervets rely on (e.g. visual, auditory) and what kinds of outputs they make to each input ("what"). It would also explain why these specific predators are the ones that elicit these calls, types and rates of errors, the degree of innateness or learning involved, and the kinds of selection pressures that might have given rise to this system ("why"). This explanation stops short of detailing the precise algorithm that is used to encode the input in a mental representation and how that representation is then translated into appropriate motor activity ("how"), let alone the neural circuitry recruited in the task. But I would add an ontological dimension to Marr's methodological and epistemic account

---

[17] On this point, I disagree with Ludwig (2016), who argues that non-epistemic factors play a role in choosing a scientific ontology.

of the computational level. Computational explanations work precisely because there are relatively self-contained causal processes in the cognitive domain that can be understood somewhat independently of the algorithms that enable them to perform their functions and the physical structures that implement them. Furthermore, computational or cognitive systems are functionally individuated, since what a system does in solving a problem or performing a task is naturally understood in terms of its function. Computational systems are a species of functional systems and computational explanations are a type of functional explanation. In this context, functions can be understood both etiologically (selected-effect functions) and synchronically (causal-role functions). Indeed, in some cases, they can be understood as combining synchronic and etiological elements.[18] The function of a cognitive system is understood both as a response to a problem in the organism's current environment, as well as, (sometimes) an adaptive response that has been naturally selected. This means that cognitive entities are individuated both environmentally and etiologically, in keeping with the argument of previous sections.

The cognitive domain is an aspect of the world that lends itself to scientific inquiry, although it cannot (obviously) be considered "mind-independent." In much recent philosophical work, "mind-independence" has been considered criterial for realism about a certain domain, but this is ill-advised, at least if we assume (as I have) that mental entities can be real. To be sure, there have been some attempts to distinguish different kinds of mind-dependence, which argue that only certain types of mind-dependence preclude realism about an entity. But I have argued elsewhere (Khalidi 2016b) that mind-independence is not an appropriate criterion by which to distinguish real from non-real kinds. Whether or not a kind depends on the mind, whether causally, constitutively, metaphysically, definitionally, or otherwise, does not have anything to do with whether it is a real kind. If we think minds themselves are real, then dependence on them should not be taken as a sign that something should not be admitted into our ontology. Rather, I would suggest that realism about kinds be understood in terms of the account of kinds that I have already outlined in this chapter (especially Section 1.2), namely in terms of being

---

[18] Griffiths (1993, 410) has proposed that "the proper functions of a biological trait are the functions it is assigned in a [causal-role] functional explanation of the fitness of the ancestral bearers of that trait." Garson (2019) has proposed a variation on the selected effect account of function that involves differential retention of traits or features in addition to differential reproduction, which may serve to capture the notion of function needed in cognitive science, whether or not evolutionary considerations are relevant. In particular, it handles cases of trial-and-error learning.

part of the causal structure of the universe. If minds, their states, capacities, processes, and so on, are causally efficacious then that is what makes them features of reality.[19] Mind-independence has nothing to do with it. To be sure, there are interesting differences between mind-dependent and mind-independent kinds, as well as between different kinds of mind-dependence, but none of these distinctions demarcate the distinction between real and non-real kinds.

It is obvious that all mental kinds, including cognitive kinds, are unavoidably mind-dependent (causally, constitutively, and otherwise), but it is debatable whether they are mind-dependent in another sense. One way to capture this additional dimension of mind-dependence is by dubbing it "response-dependence." Now there is a trivial sense in which many cognitive kinds are response-dependent. For example, whether or not I possess the concept KUMQUAT seems to depend at least in part on the responses I give to certain stimuli in my environment. Among other things, it depends on whether I can sort or categorize kumquats in the grocery store, whether I can discriminate kumquats from loquats, answer questions about their color and taste, and so on. But is possession of the concept just a fact about *my own responses* and other causal powers (as well as perhaps my causal history), or does it also depend on *others' responses* toward me? It is controversial whether some cognitive kinds are response-dependent in this second sense, as is held, for example, by an interpretivist or ascriptionist view of concepts. On such a view, whether or not a thinker possesses a concept and which concepts are possessed by a thinker are matters that depend ultimately on how they are interpreted by others; they are facts about the way in which others respond to them. In Chapter 2, I will tackle the question as to whether concepts are response-dependent in this sense, but for now I want to make the point that even if they were, this should not undermine realism about concepts (or other cognitive kinds). Many social kinds exhibit response-dependence, yet they are robust causal kinds nevertheless. As various philosophers of social science have argued, this type of mind-dependence is compatible with an entity having causal efficacy in the social domain (Hacking 1995; Mallon 2003). For example,

---

[19] Some philosophers (e.g. Kim 1992) would question the claim that minds are causally efficacious. I will not try to justify that claim here, though I have put forward some arguments against the view that causal efficacy pertains exclusively to the most fundamental entities in the universe (see Khalidi 2011). One of those arguments can be summarized succinctly as follows. Consider a possible world just like ours except that there is no fundamental level, a possibility that some scientists and philosophers take seriously and consider to be coherent (see e.g. Block 2003; Schaffer 2003). Would this be a world in which there is *no* causation? If not, it seems to be a mistake to confine causation to the most fundamental level.

social kinds like *money*, *ritual*, *gender*, and *race* are often thought to be mind-dependent in this sense. Thus, what it is to be a *woman* or *man* in many contemporary societies is thought to depend ultimately on how they are perceived and represented by others in their society (e.g. Haslanger 2000; Àsta 2013), even though these perceptions are based on how people themselves behave, are socialized, and present themselves. Yet *gender* is a real kind with robust causes and effects in the social domain. If some kinds in the psychological or cognitive domain are likewise response-dependent, then this would not seem to undermine their causal nature. Interpretivism about the mental is sometimes thought to dictate an anti-realist or instrumentalist position about mental properties or kinds, but even though this may be the position of some interpretivists and their critics, it seems possible for someone to hold both that mental entities are mind-dependent in this sense (i.e. response-dependent) yet real nevertheless.

There is a certain self-reflexivity involved in using our mental capacities to study those mental capacities themselves. Cognitive science is self-referential in a way that most other sciences are not: we are trying to understand the human mind (as well as the minds of other creatures) using the resources of the human mind. Perhaps this reflexivity is sometimes overblown, but it does lead to some tricky situations when thinking about categorization and kinds. This reflexivity comes to the fore in the very next chapter, which focuses on concepts, a topic intimately tied to the whole issue of categorization, since I take it that categories are best understood as classificatory concepts (as mentioned in Section 1.2). Thus, we will be trying to determine the nature of concepts while at the same time presupposing that there are such things as classificatory concepts that are used to pick them out. To put it more succinctly, we are attempting to articulate the concept of *concept*. But though some care needs to be taken in using the mind to study the mind itself, this inquiry does not seem to lead inevitably to a kind of paradox, or to result in hopeless subjectivism. Another way to bring out the self-reflexivity involved in the scientific study of the mind is to observe that real kinds crop up in two guises in cognitive science. The first is the one that I have emphasized so far: the identification of the real divisions or joints in the mind–brain. The second is subtly different and interacts in a complex way with the first. That is the capacity of the human mind to identify such divisions or joints in the world (including the mind–brain). Many cognitive scientists are interested in the psychological processes of categorization and in the nature of the categories that we use to understand the world and ourselves. They investigate the structure of these categories and their manner of implementation in the brain.

They are also interested in how these categories are acquired and the extent to which they are innate. This dual role for categories in cognitive science, as aspects of the theory and as elements of the mind itself, might give rise to a worry. Could it be that the categories that we use in our scientific accounts of the world are just an outgrowth of the categories inherent to our minds, which are either innate or acquired early in development on the basis of insufficient or faulty evidence?[20] This is a definite risk, but the enterprise of science aims partly to ensure that such categories, if faulty, do not persist in higher-order cognition. We have ways of weeding out categories in science if they do not play an epistemic role, and I have argued that the epistemic role of scientific categories is based on the causal role of kinds. This gives us some grounds for thinking that faulty categories can be detected and discarded. Many such mental categories have been abandoned in intellectual history and the history of science (e.g. many of the categories associated with humorism, phrenology, and psychoanalysis). But I would argue that widescale conceptual removal and replacement is rarer than philosophers sometimes suppose because successive scientific theories are generally formulated over largely the same conceptual base. Part of the justification for this assertion lies in the theory of concepts that I will defend in the next chapter. This is one respect in which there is a reflexive relationship between the theory of categorization presented in this chapter and the account of concepts put forward in the next chapter. But though this means that the theory of concepts to be proposed in the next chapter both corroborates and is supported by the account of categorization that I have defended in this chapter, I take this to be an unavoidable feature of any theory that considers concepts as a taxonomic category in cognitive science, and not as a case of vicious circularity.

1.6 Conclusion

In this chapter, I have tried to motivate a naturalist and realist account of cognitive kinds, and have argued against the reduction of cognitive kinds to neural kinds. But if, as I have argued, there are principled obstacles to reducing our mental categories to neuroscientific ones, does that mean

---

[20] For example, Leslie (2013, 109) argues that essentialist positions in philosophy rest on intuitions that are "due to a deep-seated cognitive bias, rather than to any special insight into the nature of reality." But the fact that such positions are not universally accepted and have been widely criticized in both philosophy and cognitive science is evidence that we are not fated to adopt our intuitive and innate categories.

that neuroscience cannot shed light on our mental lives? Even though there is not likely to be a one-to-one correspondence between the mental and neural in all cases, there will be many ways in which a better understanding of neural mechanisms may help shed light on cognitive processes. Discoveries in neuroscience can certainly inform our psychological theories and explanations even though they are not likely to replace or preempt them. The relationship may be compared to that between genotype and phenotype, which is also not a reductive one, at least according to many philosophers of biology (see e.g. Kitcher 1984; Schaffner 1998; Wimsatt 2007). Once it was thought that genes would correspond directly to phenotypic features and that we would be able to read off the phenotype from the genotype. But as we find out more about the complex relationship between the two, it has become clear that this is just not the case, even when it comes to fairly simple traits like eye color and height in humans.[21] A vanishingly small number of phenotypic traits can be traced directly to single allelic variants or even a limited number of them. It may be objected here that when it comes to genotype and phenotype, the former causes the latter and does so in conjunction with other causes, but neurophysiological processes do not cause mental ones, since they are identical with them. Even though some instances of neurophysiological tokens and types may be identical with cognitive tokens or types, or bear a compositional or mereological relation to them, I do not think this holds generally, as I argued in Section 1.4. In most cases, when we individuate the items under consideration carefully, the relationship between the neural and cognitive turns out not to be one of identity, either when it comes to types or tokens. The relationship may not always be straightforwardly causal either, but it combines elements of causation, composition, constitution, and other relations, as I will try to indicate in later chapters.

---

[21] According to current estimates, there are around 20,000 genetic variations (i.e. single nucleotide polymorphisms (SNPs), out of a total 4–5 million, i.e. 0.5 percent) in the human genome that are thought to influence a person's height (Lello, Avery, Tellier, et al. 2018).

CHAPTER 2

Concepts

I have a young conception in my brain;
Be you my time to bring it to some shape.
— William Shakespeare, *Troilus and Cressida*

Every new concept first comes to the mind in a judgment.
— C. S. Peirce, "Belief and Judgment"

2.1 Introduction

Concepts are often considered the most basic element in our cognitive ontology; they are frequently regarded as the building blocks of the mind or the vehicles of thought. The past few decades have seen a surge of work when it comes to concepts, both theoretical and empirical. Once the exclusive province of philosophers, concepts have become the stomping ground of cognitive scientists over the past several decades. There are currently a number of different research programs investigating concepts empirically using a variety of methods. Even though this is seldom made explicit, I would argue that they do not always address the same questions. There are at least five different questions that are commonly at issue in recent theoretical and empirical work on concepts. First, what types of entities are concepts? Can they be *identified* with mental entities, neural entities, both, or neither? Are they concrete particulars or abstract entities of a certain kind? Second, there is a question about concept *individuation* or grounding: In virtue of what does a concept have the content that it does? In other words, what gives a concept its identity conditions, or what makes something the concept of APPLE as opposed to ORANGE? Third, researchers are interested in concept *acquisition*: How are concepts acquired in ontogeny? Are some concepts innate, or are they all learned, and by what processes are concepts learned? (There is also some research concerning the acquisition of concepts in phylogeny, though for obvious reasons there is less empirical work relevant to that question.) Fourth, there is a question about concept *possession*: What is it for a thinker to have a concept, or what determines whether a subject has acquired a concept? Finally, there is the issue of concept *activation*: What is it for a concept to be activated in the mind on a particular occasion (which is roughly the same as conceptual retrieval or processing)? What goes on in the mind or brain of thinkers when they entertain a particular concept? To anticipate, I will argue in due course that some research programs are more focused on some of these questions explananda than others. Hence, they are sometimes working at cross purposes and are not attempting to answer the same questions or explain the same phenomena.

In addition to the fact that there are at least five different research questions or explananda that the research on concepts addresses, there are a number of theoretical controversies that split researchers on concepts into various camps. These theoretical controversies do not map neatly onto the explananda mentioned above, in the sense that some of the theories provide an answer to more than one of the following questions or attempt to explain more than one of these phenomena (without always clearly distinguishing them). What follows is not an exhaustive list of debates about concepts, but these theoretical disputes are among the most prominent ones and they are the ones that will figure in the discussion to follow:

1) ***Externalism vs. Internalism (Individualism)***: An "externalist" view of concepts holds that concepts are individuated by determinants that are external to the mind of the agent, while an "internalist" (or "individualist") position attends to the subject's perspective on the world in individuating concepts.

2) ***Modal (Sensorimotor) vs. Amodal Theories***: This is a distinction, which is grounded in some traditional philosophical debates as well as in the recent empirical literature, between accounts of concepts that are based in our sensory modalities or sensorimotor abilities and those that are amodal (or trans-modal). The former regard concepts, whether concrete or abstract, to have their contents as a result of their connections to sensory, affective, and motor functions or processes, whereas the latter do not think that sensory, affective, and motor abilities can account for the contents of concepts.

3) ***Definitional (Classical) vs. Prototype vs. Theory Theories***: Another debate is that between those who take the content of concepts to be supplied by a definition or consider them to be structured in terms of

necessary and sufficient conditions, and those who deny this, judging the content of a concept to consist of a prototype of weighted features with a family resemblance structure. There are also other accounts of conceptual content or structure, most notably, those accounts that take concepts to be embedded in theories and inextricable from them.

4) ***Holism vs. Atomism***: This debate concerns whether the content of a concept depends in systematic ways on the contents of the other concepts in a conceptual repertoire ("holism"), or whether each concept has its content independently of others ("atomism"). One can also distinguish an intermediate position that considers the content of each concept to depend only on *some* other concepts ("molecularism").

5) ***Response-Dependence vs. -Independence***: This is a dispute between those who consider the identity conditions for concepts to be dependent ultimately on human responses and judgments, and those who do not hold them to be response-dependent in this way. If concepts are response-dependent, that means that they are phenomena whose very identity is determined by the judgment of observers or interpreters (this is the case for an "interpretive" or "ascriptive" view of concepts). If they are not, then the content of a concept or its individuation conditions are not dependent on the response of an interpreter or the judgment of the community.

6) ***Minimalism vs. Intellectualism (Maximalism)***: There is a distinction that is sometimes made between "intellectualist" theories of concepts, which consider that language and higher cognitive abilities are required for the possession of concepts, and "minimalist" theories of concepts, which hold that they merely involve discriminatory abilities or sensitivities along with certain combinatorial abilities. The latter account of concepts is supposed to countenance the possession of concepts by prelinguistic infants and nonhuman animals, whereas the former is thought to restrict concept possession to language-using creatures.

These would seem to be the most prominent points of disagreement among those researchers – philosophers, linguists, psychologists, neuroscientists, and others – who are engaged in inquiring into the nature and structure of concepts. In what follows, I will be touching on each of these controversies to various degrees and will in some cases propose ways of reconciling the different approaches or at least rendering them compatible. To illustrate the point that these theoretical approaches do not correspond neatly to the earlier research questions outlined earlier, consider the first theoretical dispute between externalists and internalists. This is primarily a disagreement concerning the question about conceptual individuation, since these theorists give different answers to the question as to what gives concepts their contents or semantic values. But these theorists also tend to differ when it comes to the second question concerning acquisition, since many externalists think that concepts are acquired by thinkers (roughly) when they are causally connected in some way to the referents of those concepts, whereas internalists tend to think that thinkers can acquire concepts by relating them to other concepts in the right ways. This also leads to a difference of opinion about conditions on concept possession. Similar points apply to some of the other theoretical debates mentioned in the previous paragraph, which have repercussions for more than one of the questions mentioned: identification, individuation, acquisition, possession, and activation.

The rigorous empirical study of concepts has only been around for roughly half a century, but scientists have devised various methodological paradigms to ascertain concept activation, as well as concept possession and acquisition. There is a wide range of methods and tasks used, many of them involving responding to verbal stimuli and images that represent, illustrate, or are otherwise associated with particular concepts. The concepts investigated are predominantly those denoted by common nouns for concrete objects (e.g. "apple," "chair," "fruit," "furniture"), though concepts denoted by verbs are occasionally also studied. The tasks usually involve categorization (including assent or dissent from statements concerning categorization, e.g. "an apple is a fruit"), recognition, discrimination, and inference. When experimenters rely on verbal and behavioral responses, these responses are sometimes elicited under time constraints or while measuring reaction times. In some cases, there are no such constraints and participants answer at their leisure and are asked to justify their responses. Finally, in addition to verbal and behavioral responses, studies sometimes involve neural ones, measured either by the blood-oxygen-level-dependent (BOLD) signal in a fMRI scan or the electrophysiological reading indicated by an electroencephalogram (EEG). Though there are many experimental paradigms that are not covered by this brief overview, these are some of the typical methods used in empirical research on concepts, and we will encounter them again in what follows. In Section 2.2, I will present some research on concepts in cognitive neuroscience, and introduce two theoretical approaches that have emerged from this work, which relies primarily on neural evidence. Then, in Section 2.3, I will survey some of the main results in research on concepts in cognitive psychology, which depends on evidence from behavioral and verbal responses, as well as the main theoretical constructs. After that, in Section 2.4, I will present a functional theory of concepts that has some affinity with one theoretical approach in cognitive psychology, arguing that other theories and taxonomic categories in psychology and neuroscience may be tracking different cognitive processes and identifying different kinds. In Section 2.5, I will consider and reply to a number of objections to this theory of concepts, before concluding in Section 2.6.

2.2 Empirical Accounts: Cognitive Neuroscience

Most cognitive scientists identify concepts with types of mental representation, and for many, mental representations are supposed to be implemented in certain types of neural entities or processes (e.g. populations of neurons, patterns of neural activations), though few if any cognitive scientists would claim to know exactly how they are so implemented at present. Many researchers on concepts, especially in cognitive neuroscience, are therefore interested in isolating the neural correlates of concept activation, paving the way for a reductionist account of what it is for a thinker to entertain a concept on a particular occasion. Some of these theories also contain or lend themselves to a particular account of how concepts get their contents, or what it is about a particular pattern of neural activity that constitutes the activation of a particular concept. They are hence theories of individuation or grounding as well as activation. They can also be seen to give an account of possession, though this is not usually their focus, as I will go on to argue. In this section, I will present what is currently the dominant theory of concept activation and individuation in cognitive neuroscience, which closely associates conceptual representations with perceptual ones. (For brevity, I will sometimes talk exclusively of perceptual or sensory systems, but motor and affective systems should also be understood to be implicated in the activation of at least some concepts, on this view.) After outlining what I take to be some of the main problems with this theory, I will then sketch an alternative picture that has been proposed by some cognitive neuroscientists, which problematizes the search for concepts in neural infrastructure.

On a modal theory of concepts the deployment of a concept on a particular occasion consists in the occurrence of patterns of neural activation that reactivate states in modality-specific systems in the brain, namely perceptual, motor, and affective systems. On one version of this theory, the activation of a concept consists in large part of the "re-enactment" of sensorimotor perceptual representations associated with that particular concept, where a reenactment "partially reproduces experienced states" (Barsalou, Simmons, Barbey, et al. 2003, 88). Since the mere reenactment of perceptual features would not seem to account for full-fledged conceptual abilities, including the ability to categorize and infer, many accounts also invoke additional neural processes. On most modal views, it is not enough for these neural populations to be coactivated to be bound together in a single concept, they have to be conjoined in some way, and on some, conjunctive neurons are responsible for binding and reactivating neuronal populations responsible for encoding perceptual features. Hence, the reenactment consists not just in the reactivation of perceptual representations but also in the activation of "cross-modal" neurons in "convergence zones," which are also known as "association areas" or "hubs" in the brain (Barsalou 2016, 1132). These convergence zones are supposed to be responsible for binding the percepts together and are posited to integrate information across modalities. Thus, among modal theorists, many claim that all that is required for the representation of even abstract concepts is perceptual representation along with certain conjunctive operations carried out by cross-modal neural systems, and they think of "abstract conceptual representations as high-level conjunctions rather than amodal symbols" (Binder 2016, 1098). While some modal theorists occasionally invoke "amodal" representations, this would seem to undermine the theory's most basic tenet that conceptual representation is fundamentally modally grounded. Moreover, as some have pointed out, "amodal" as used by modal theorists should really be taken to mean "multimodal" (a term often used interchangeably with "crossmodal" and "transmodal," or occasionally "supramodal"), and is used to describe representations that somehow combine or bind representational content from multiple sensory modalities (Barsalou 2016, 1126; Binder 2016, 1098; Spunt & Lieberman 2012; cf. Kemmerer 2019, 41 n.1). This account is supposed to apply to abstract as well as concrete concepts, and indeed erase the distinction between the two types of concept (Barsalou, Dutriaux, & Scheepers 2018).

Some evidence for modal theories of concepts comes from experiments that exclusively rely on behvioral techniques and measures, such as reaction times, error rates, and frequency of listing certain features. For instance, as reviewed by Barsalou, Simmons, Barbey, et al. (2003), when participants are asked to list features associated with a concept, features are reported less if they are typically occluded perceptually (e.g. participants produce "roots" less often for "lawn" than for "rolled-up lawn"), and when participants are asked to verify sensory features associated with certain concepts they are slower if they had been asked to verify a feature from a different modality on a preceding trial (e.g. verifying "loud" for blender is faster after previously verifying "rustling" for leaves than after verifying "tart" for cranberries).

2.2 Empirical Accounts: Cognitive Neuroscience

The experimental design presupposes that if conceptual representations are modality-specific, switching from one modality to another should slow verification. But since these are at best indirect indications of the underlying structure or neural organization of conceptual representations, more compelling evidence for modal theories of concepts comes from recently developed experimental techniques in cognitive neuroscience. Though the tasks in these experiments are primarily standard behavioral tasks involving categorization or recognition, the main methodological development is the use of neuroimaging technology to measure neural activity while participants undertake such tasks. The variety of experimental protocols that have been used in this area of research is difficult to summarize here, but most experiments require experimental participants to perform certain cognitive tasks relating to concepts while undergoing a fMRI scan. In a standard experimental paradigm, participants are simply asked to read individual words in the scanner while their hemodynamic activity is monitored. In one highly cited study, experimenters selected a number of action words associated with face, arm, and leg movements (e.g. "lick," "pick," "kick"). Participants in the experiment were asked to read individual words on a screen while fMRI was used to gauge levels of neural activation in their brains (Hauk, Johnsrude, & Pulvermüller 2004). The hemodynamic response when reading these words was then compared to the response when participants were asked to move their left or right foot, left or right index finger, or tongue. In this instance, experimenters found that the action words differentially activated areas along the motor strip that were either directly adjacent to or overlapped with areas activated while moving the corresponding body parts, thereby providing support for a modal theory of concept representation. In such an experimental paradigm, it is difficult to ensure that subjects are indeed focusing on the conceptual content of the words they are reading and that they are not simultaneously distracted by other thoughts. Indeed, for any given task, it is problematic to determine which precise concepts are being activated, since there is seldom just a single concept at play in any given cognitive task, even under controlled experimental conditions. Thus, to avoid confounds, various attempts have been made to design tasks that would serve to zero in on the concept or concepts of interest in any given condition. In a more recent experimental design, a written word denoting a concept was presented on the screen for several seconds and participants in the scanner were asked to "think deeply" about its meaning in such a way that they could determine subsequently whether it applied to a visual scene (Wilson-Mendenhall, Simmons, Martin, et al. 2013). The visual scene was then presented and participants were asked to judge whether or not the word applied to that scene. In some trials, the relevant words were not followed by a visual scene, and these "catch" trials supposedly enable researchers to separate the neural activity involved in thinking "deeply" about the concept from that involved in judging whether or not it applies to a visual scene. This way, researchers are supposed to capture the neural activation associated with deploying a concept. But despite the ingenuity of this method, there are various questions that arise even when it comes to this careful attempt to capture the activation associated with a specific concept and separate it from other cognitive processes. Perhaps most importantly, it is far from clear that asking people to explicitly think about or reflect on a concept is the best way of eliciting that concept. If one of the main functions of concepts is categorization, then the process of applying a concept to a visual scene might actually be a more valid way of gauging the neural activation associated with conceptual processing or deployment.[1]

The behavioral and neural evidence for modal theories of concepts or concept activation is compelling, but it raises a number of questions that should lead us to exercise caution. First, as already mentioned, many, if not most, concepts do not seem to be mere concatenations of perceptual features, even concrete concepts like APPLE. At the very least, modal theories need to say far more about how convergence zones enable the representation not just of simple conjunctive relations (e.g. RED and ROUND), but more complex logical and statistical ones (e.g. if RED then probably RIPE), let alone the representation of concepts that do not seem amenable to being constructed simply out of clusters of perceptual features (e.g. RIPE). Even if we allow that conjunctive neurons or convergence zones manage to bind perceptual representations in the logical relationship of conjunction, no explanation is forthcoming of other Boolean and non-Boolean aspects of conceptual structure (e.g. disjunction, if-then, part-whole, set-subset, predication, probabilistic, generic, and so on). But this would require drawing on hitherto unspecified non-perceptual representational resources. In a critique of modal theories, Weiskopf (2007, 174) has argued forcefully that when it comes to the representational properties of at least some concepts, "we will need to appeal to many neural activation patterns

---

[1] Another problem with this experimental design, as well as many others in this area, is that they rely to some extent on the "method of subtraction," which attempts to isolate the neural correlate of some cognitive process C_i by starting with some task T, which is thought to involve various cognitive processes, $C_1, C_2, \ldots C_n$, and then subtracting from the neural correlate of T the neural correlates of the other processes, $C_2, \ldots C_n$, as revealed in other experiments. But as many researchers have pointed out, this problematically assumes a simple additive relationship between cognitive processes (see e.g. Poldrack & Yarkoni 2016).

beyond those in the perceptual systems; these neural activation patterns are not themselves perceptual representations or copies thereof ..." At least as they stand, modal theories do not seem to account for a panoply of conceptual abilities that transcend perceptual (as well as motor and affective) features and include theoretical information required in making complex inferences about such matters as causality and intention. Moreover, this problem implies that modal theories of concepts do not have an obvious advantage when it comes to giving an account of concept individuation or grounding. One of the main selling points of modal theories would seem to be their ability to provide such an account, simply because conceptual representations can be said to inherit their representational content from perceptual ones, which are presumably grounded by being connected in the appropriate way to sensorimotor receptors. However, given that modal theories owe us an account of "missing" representational content, they must also provide an alternative way of accounting for grounding at least some non-perceptual aspects of conceptual content.[2]

Second, despite the existence of significant empirical results that appear to confirm modal theories of conceptual activation, there is a substantial body of counter-evidence, which has been detailed by the theory's critics. Some of this evidence consists in the existence of double dissociations among patients with neural lesions. For example, in the color domain, patients with lesions can be impaired in their color knowledge (e.g. "this is the color yellow") but can nevertheless retain knowledge of the typical colors of objects (e.g. "bananas are typically yellow"). Conversely, lesion patients can have intact color perception but impaired conceptual knowledge of the typical colors of objects (see Mahon & Hickok 2016, 949, and references therein). Numerous examples of such dissociations exist between sensorimotor capacities and conceptual capacities, indicating that it would be too hasty to conclude that conceptual representations are simply reenactments of perceptual representations, or even that they involve such reenactments as a necessary element. There is additional neuropsychological evidence against modal theories from "semantic dementia," which results in a "modality-general, item-specific" pattern of deficit, with "no apparent interaction between conceptual category and perceptual modality" (for a review and analysis, see McCaffrey 2015a, 343).

Third, modal theories have a problem when it comes to distinguishing the conceptual activation pertaining to a concept proper from activation

---

[2] Barsaolu, Dutriaux, and Scheepers (2018) seem implicitly to acknowledge this by attempting to draw on the resources of situated cognition to flesh out the representational content of concepts.

associated with a concept but not strictly proprietary to the concept itself. In this vein, Mahon and Hicock (2016, 947) point out that at least two inferences can be drawn from empirical evidence indicating that sensorimotor representations are activated during conceptual processing. The first is that concepts are represented in a sensorimotor format and activation of sensorimotor representations reflects conceptual access. The second is that concepts are represented in an amodal format and that activation spreads from these amodal representations to connected sensorimotor representations, whose activation is evidence of information flow from one system to the other. The latter inference is effectively an alternative hypothesis to explain some of the same evidence that is usually taken to support the modal theory. To rule out this hypothesis, one would have to have a principled way of distinguishing the neural activation constituting activation of the concept proper from neural activation that is associated with the concept but not constitutive of it. This poses both technical and theoretical challenges. Technically, the temporal resolution of many of our current neuroimaging tools may not be sufficiently fine-grained to distinguish the relevant cognitive processes. Conceptual processing in many experimental tasks can take place in the space of a few hundred milliseconds, whereas the temporal resolution of fMRI scans is currently of the order of one thousand milliseconds. Theoretically, it is notoriously difficult to use "functional connectivity," which is a measure of statistical correlations between regions as revealed in fMRI scans, to infer "effective connectivity," which is a measure of how regions actually interact causally. Perhaps more importantly, as I will go on to detail, there is mounting evidence to suggest that there may not be a clear theoretical distinction to be made between the activation associated with the concept itself and at least some types of associated activation. Much recent work on conceptual processing has pointed to the variability in neural activation associated with a given concept, especially as it occurs in different contexts. According to this body of research, there is no "core" of neural activation associated with any given concept across different contexts, including experimental tasks, individuals, languages, and cultures, among others. In the rest of this section, I will briefly outline this body of work and extract its relevance for the attempt to identify the neural correlates of conceptual thought.

Another major trend in recent empirical work on concepts in cognitive neuroscience emphasizes the fluidity, flexibility, and instability associated with conceptual content. The main findings in this body of research concern the inter and intrapersonal variability in neural activation when it comes to tasks involving the deployment of concepts, such as categorization and recognition. Yee and Thompson-Schill (2016) review a number of neuroimaging studies showing that neural activation varies along a variety of contexts or dimensions when it comes to conceptual tasks: long-term experience, recent experience, concurrent context or ongoing goals, and passage of time as object recognition unfolds. For example, when it comes to long-term experience, in professional musicians, identifying pictures of musical instruments activates auditory association cortex and adjacent areas more than identifying pictures of other objects, but this difference in activation does not appear in musical lay people (Hoenig, Müller, Herrnberger, et al. 2011). In such cases, modal theorists might just say that experts and nonexperts simply have different concepts, but similar results obtain for categorization and recognition tasks when it comes to the other conditions mentioned, such as a person's recent experience. For example, even though many neuroimaging experiments confirm that words relating to actions recruit the motor system and words relating to colors recruit visual areas, some studies also demonstrate that within motor areas neural activation levels are higher when participants are instructed to focus on the action associated with an object word than when they are instructed to focus on the object's color (van Dam, van Dijk, Bekkering, et al. 2012). Similarly, based on neuroimaging evidence, Hoenig, Sim, Bochev, et al. (2008, 1809) concluded that processing of a particular concept "does not selectively and constantly activate a specific sensory or motor region," since neural activation is based on contextual constraints, specifically the way in which the concept is cued, whether by using visual attributes or action attributes. Moreover, they used event-related potentials (ERPs) from electroencephalogram (EEG) experiments to rule out the possibility that this pattern of activation was a result of top-down modulation of conceptual processing, since the effects in question were present within 200 milliseconds of target onset (cf. van Dam, van Dijk, Bekkering, et al. 2012).[3] Collectively, such studies have led many researchers to conclude that neural activation in conceptual tasks is highly context-dependent and varies not just from person to person but also from occasion to occasion, depending on various contextual factors. This puts pressure on the attempt

---

[3] Both Hoenig, Sim, Bochev, et al. (2008) and van Dam, van Dijk, Bekkering, et al. (2012) use EEG to measure conceptual processing and claim that context-dependent effects pertain to conceptual processing proper since they occur within 200 milliseconds of stimulus onset. McCaffrey and Machery (2012, 273) dispute that researchers have a principled way of knowing that such effects are instances of conceptual processing even if they occur in such a short period of time after stimulus onset. As will emerge in due course, I think that such problems are just symptoms of the inherent difficulty involved in attempting to identify concepts with patterns of neural activations.

to identify concepts with determinate neural correlates that are relatively constant in each individual, let alone across individuals.

This recent research builds on an older body of work that relies on behavioral evidence, which finds that participants' categorizations often vary with respect to context and are out of step with their own considered judgments, sometimes without warrant. In many classic studies on categorization, researchers found that the features associated with a given concept vary widely across experimental contexts. As we might expect, different features are associated with the same concept in different contexts (e.g. with the concept PIANO in the context of producing music and moving furniture; see Barclay, Bransford, Franks, et al. 1974), and many associated features in property verification tasks appear and disappear with context (e.g. the feature "has lungs" is associated with the concept BEAR in the context of the sentence, "The bear caught pneumonia," but not in other contexts; see Barsalou 1982). More drastically, psychologists have found that people's considered judgments about concept membership do not always accord with their categorizations of various instances on particular occasions. Among numerous other results, participants in some experimental conditions judge some odd numbers to be "more odd" than others and rank them on a scale of "oddness," even though they also agree in other conditions that the concept ODD NUMBER is all-or-none (Gleitman, Armstrong, & Gleitman 1983; cf. Barsalou 1983). Thus, just as the pattern of neural activations associated with a concept can vary considerably across contexts, the constituent features associated by an individual with a particular concept are also not constant in the different contexts in which that concept is accessed.

These experimental results suggest that there is good reason to doubt that a pattern of neural activation that is found to be associated with a concept in a certain experimental task is *constitutive* of that very concept, as opposed to merely being associated with it in some way. The variability in the way that concepts are manifested in different contexts cautions against equating neural activation on a given occasion with the manifestation of the relevant concept. However, those researchers who would identify the manifestation of a concept with some determinate pattern of neural activation, particularly in sensorimotor areas, could respond by saying either that: (a) there might yet be a conceptual "core" activation pattern common to all contexts; or (b) the concept might be identified with the *totality* of activations associated with the concept; or (c) there may exist some higher-dimensional invariance despite the apparent variability, and we might be able to extract some kind of constancy amidst the flux in the neural activation data. Each of these possibilities must be considered separately. When

2.2 Empirical Accounts: Cognitive Neuroscience

it comes to (a), some of the neuroimaging evidence already cited strongly indicates that there is no core common to the processing of many concepts. After surveying a body of research that relies on both behavioral and neuroimaging evidence, Lebois, Wilson-Mendenhall, and Barsalou (2015, 1772) argue for a "no-core" theory of concepts, concluding: "Rather than concepts containing cores that are activated automatically independent of context, concepts only appear to contain information that is dynamic and context-dependent." Similarly, Hoenig, Sim, Bochev, et al. (2008, 1801) write that there is a need to posit a strong form of conceptual flexibility "in order to account for the high degree of heterogeneity in category-related imaging findings with no single study detecting all the hitherto reported category-related activation foci …, and some studies even failing to find any category-related effects …" Some of this evidence counts not just against modal theories, but amodal ones as well. As Lebois, Wilson-Mendenhall, and Barsalou (2015, 1773) state: "the evidence we have reviewed suggests that concepts have no cores at all, amodal or modal." As for (b), identifying the concept proper with the totality of all such neural activations, this is problematic especially if there is no core to conceptual representations, since these activations will then be disjoint. A set of disjoint neural activations is a poor candidate for the neural correlate of a conceptual representation, at least if the aim is to effect a reduction of concept deployment and possession in neural terms. To put it in familiar philosophical terms, this would mean that concepts are multiply realized by their neural correlates. Finally, when it comes to (c), the possibility of discovering some higher-dimensional commonality in the variable patterns of activation, this is at present just a promissory note. It depends on the discovery of some nonobvious higher-order function that would serve to unite the disjoint neural correlates. Moreover, it would not seem to account for the variability in behavioral and verbal responses, which also points in the direction of fluidity and instability.[4]

Increasingly, many of the researchers involved in generating and interpreting the relevant data about the neural correlates of concepts insist that the correct interpretation of these results is the radical one that there is no stable, self-contained, context-independent neural representation that corresponds to any given concept. As Yee and Thompson-Schill (2016, 1016; original emphasis) put it: "… *the concepts themselves are inextricably*

---

[4] As will become clearer in the course of this chapter, I think it is a mistake to expect a single theory to explain fluidity and instability at the neural and behavioral levels, but researchers in this area often aim at a reductive account.

linked to the contexts in which they appear, so much so that the dividing line between a concept and a context may be impossible to clearly make out ..." These researchers also maintain that one should think of "the concept itself as changing slightly each time it is retrieved, and that there is no real demarcation between what is activated in a given instance and the concept itself" (2016, 1018). Connell and Lynott (2014, 393) perhaps state it most radically and most pithily: "one cannot, in effect, ever represent the same concept twice." Thus, a large body of recent research leads to the conclusion that the neural correlates of concept activation are highly variable and contextually determined.

2.3 Empirical Accounts: Cognitive Psychology

In cognitive psychology, two main theories of concepts have been dominant in the past few decades: (i) prototype theory, which holds that the content of a concept is based on a weighted set of features or stored exemplars, and (ii) theory theory, which posits that the content of a concept is based on its links to other concepts or its place in an encompassing theory. In this section, I will attempt to outline both theories and briefly introduce some of the empirical evidence that supports each. I will also try to relate these theories to those encountered in the previous section. Then, I will consider two ways of reconciling the two theories, and give some reasons to favor one over the other, which will allow me to introduce a different approach to concepts in Section 2.4. But before presenting each of these theories, I will briefly review the "classical" or "definitional" view of concepts that these two more recent views are usually regarded as displacing.

On a classical view of concepts, a concept consists in a definition, which supplies necessary and sufficient conditions for falling under the concept in question. To possess a concept is to have the definition in mind and to be able to deploy it in cognitive tasks such as categorization and inference. Moreover, to have the concept in mind or to activate it on a given occasion is to entertain the definition, and acquiring the concept consists in coming to have the definition. This view of concepts has been largely abandoned by philosophers and (especially) psychologists, for a variety of reasons, which are worth recounting very briefly. First, to say that concepts consist in and are grounded by their definitions seems to pass the buck, since those definitions presumably themselves consist in concepts (e.g. an apple is a fruit that is round, sweet when ripe, etc.). The natural way to stop this regress is to say that it bottoms out in a set of conceptual primitives, often thought to consist in a set of percepts and logical connectives. The percepts are then grounded in their connections to the sources of the relevant perceptual stimuli. But it is generally agreed that the empiricist philosophical project of building up concepts from percepts has largely failed and that our conceptual edifice cannot be constructed in this systematic fashion from percepts (cf. Fodor 1981). Another problem is that very few if any concepts are amenable to strict definitions. Any set of necessary and sufficient conditions that are put forward for defining even seemingly innocent and straightforward concepts are bedevilled by exceptions, and in numerous cases no amount of tinkering manages to avoid the problem (see e.g. Fodor 1981, 283–292). Third, even concepts that seem amenable to definitions at a particular stage of inquiry do not always retain those definitions as the inquiry progresses. This occurs time and again in intellectual history and the history of science, even though the concepts in question appear to persist, which suggests that definitions do not ground or provide identity conditions for concepts. Although the problem might be partly addressed by equating concepts with their definitions as determined *at the end of inquiry*, this would force us to admit that almost all our concepts are discarded and replaced with each definitional adjustment over the course of intellectual history. The same would hold for individual cognitive development, since the definitions associated with our concepts are often modified in the course of ontogeny, which would give rise to widespread conceptual replacement.[5] A fourth problem pertains specifically to concept possession: We are often warranted in attributing possession of a concept to a thinker even when that thinker is unaware of the purported definition, whether implicitly or explicitly (e.g. a schoolchild may have the concept CIRCLE despite not knowing that a circle is a plane figure all of whose points are equidistant from a given point). A fifth problem relates mainly to conceptual activation: Some empirical work indicates that entertaining a concept does not generally involve entertaining the concepts that occur in its definition (see e.g. Fodor, Garrett, Walker, et al. 1980). Moreover, other empirical results show that concepts are structured not in terms of definitions but in terms of features that are neither singly necessary nor jointly sufficient for concept membership, as we will see shortly in recapitulating some of the evidence in favor of the prototype theory. These are some of the main problems that have led philosophers and cognitive scientists to

---

[5] Some cognitive psychologists are willing to accept such a consequence (e.g. Gopnik 1984; 1988; Carey 1985; 1988; 2009), but I have argued elsewhere that they need not and ought not (see Khalidi 1998).

conclude that concepts should not be equated with definitions and have compelled them to look for other theories of concepts.

The central claim of prototype theory is that concepts are structured not as definitions but as sets of representations of weighted features (see e.g. Smith & Medin 1981, 61–101). For example, the concept APPLE would be a list of feature representations (e.g. ROUND, EDIBLE, SWEET), each feature being weighted according to its importance to the concept. The more features an instance shares with the concept and the more important those features, the more typical that instance is of the relevant concept. Individual thinkers recognize and categorize objects as apples based on their having attained a certain requisite "score" that takes into account the weighting of each feature, not on the basis of having a set of necessary and sufficient features. More typical exemplars have more features, or more of the important features, and would attain a higher score than less typical exemplars.

The experimental evidence that led to the development of the prototype theory relies on behavioral rather than neural measures, including explicit judgments in categorization tasks, accuracy of responses in recognition tasks, and reaction times in both categorization and recognition tasks. The theory proceeds from the familiar idea that when it comes to many categories or concepts, some instances are generally considered more central, representative, or typical. For instance, at least among many North Americans[6], an apple is considered a more typical fruit than a coconut, a robin is regarded as a more typical bird than an ostrich, and a sofa is taken to be a more typical piece of furniture than an ottoman. This familiar fact is taken not just to be a matter of conventional wisdom but to reflect the way in which concepts are represented in the minds of individual thinkers. There are several different types of experimental findings that have been deemed to be evidence for typicality effects. One type of experimental result finds that for many concepts or categories, some instances are considered more typical by most participants when they are explicitly asked to rate them on a typicality scale. For example, participants are asked to rate the extent to which an instance represents their "idea or image of the category" (Rosch 1975, 199; cf. Rosch & Mervis 1975, 588). They are given words such as "orange," "apple," "pineapple," and "coconut," and are asked to rate their typicality as instances of the concept FRUIT on a scale from 1 to 7. Unsurprisingly, "orange" is rated more typical than "pineapple," which

---

[6] It is fairly uncontroversial that specific typicality results are culturally variable, but there is also some debate as to whether typicality results are found at all in some cultures; for discussion see Kemmerer (2019, 67–68) and references therein.

is rated more typical than "coconut," with a high degree of agreement among participants (undergraduates living in California) (Rosch 1975, 198). The results of this experiment demonstrate that some instances of a concept are generally regarded as more typical than others, but so far, this does not necessarily reveal anything about the structure of conceptual representations, since it may just reflect conventional or common knowledge about typicality. A second type of experimental result aims to establish that the instances of a concept that participants consider more typical are also those that are found to share more features with other instances of that concept. In this experiment, participants are given words such as "orange," "apple," "pineapple," and "coconut," along with the instruction to list as many features of that fruit as possible (e.g. "sweet," "juicy," "round," and so on) within 90 seconds. The fruits judged more typical (e.g. orange) are found to share more features with other fruits than the ones judged less typical (e.g. coconut) (Rosch & Mervis 1975, 582). A third type of result finds that participants are faster when categorizing those instances that are regarded as more typical. For example, participants are asked to respond "true" or "false" to such statements as "An apple is a fruit," "A pineapple is a fruit," and "A coconut is a fruit," and their reaction times and error rates are measured. Shorter reaction times are recorded for sentences involving the instances judged more typical (Rosch 1978, 38).[7] A fourth experimental result finds that superordinate concepts are better primes for typical instances than nontypical instances. When participants are asked to judge quickly whether two simultaneously presented words or images are identical, the word for the superordinate concept acts as a prime only for word pairs naming typical instances (Rosch 1975). As the experimenters put it: "Apparently, hearing the category name leads subjects to expect typical category items and not to expect atypical items" (Rosch, Simpson, & Miller 1976, 498). Finally, prototype theorists claim that the words for typical instances are likely to be named first and more frequently when subjects are asked to list instances of a certain concept, and the words for typical instances are the first ones to be learned by children and are learned more quickly by them (Rosch 1978, 38–39). Such findings are meant to support the prototype theory in various ways. For example, a concept instance that has more features or more heavily weighted features

---

[7] In the same series of experiments, participants were instructed that "some reds are redder than others" and that a Pekinese is a "less doggy dog" than a Retriever or a German Shepherd. The instructions also read, in part: "Don't worry about *why* you feel that something is or isn't a good example of the category ... Just mark it the way you see it" (Rosch & Mervis 1975, 589; emphasis added).

will be categorized more rapidly due to the fact that it achieves the requisite "score" in a shorter amount of time. Thus, typicality judgments are thought to reflect something about conceptual structure and the nature of conceptual representation.

The main competitor of the prototype theory is the theory theory of concepts, which has not been modeled as precisely as the prototype theory, but has been put forward as an alternative picture of conceptual structure.[8] According to the theory theory of concepts, concepts are embedded in a larger framework of explanatory beliefs (or theories), which thinkers draw upon in performing a particular cognitive task. Different parts of the entire corpus of beliefs may be deployed in different tasks, even ones involving a single concept. Proponents of the theory theory posit an interrelated network of conceptual information rather than self-contained collections of feature lists. Since the content of a concept depends in systematic ways on the contents of other concepts in a conceptual repertoire, this would make the theory theory a holist rather than an atomist account of concepts. As Murphy and Medin (1985, 289) once put it in a seminal article, referring to the then-dominant prototype theory:

> [C]urrent ideas, maxims, and theories concerning the structure of concepts ... are inadequate, in part, because they fail to represent intra- and inter-concept relations and more general world knowledge. We propose a different approach in which attention is focused on people's theories about the world ...

They go on to say: "we wish to reduce the importance of individual attributes [i.e. features] in conceptual representations and to emphasize the interaction of concepts in theory-like mental structures" (Murphy & Medin 1985, 292). This alternative picture is based on experimental evidence indicating that in many categorization tasks, thinkers rely not (or not just) on typical features or features that are statistically or probabilistically associated with concepts. Rather, they draw on theoretical, explanatory, and causal information related to the concept. Especially in non-routine categorization and inference, experimental participants forego the former type of information in favor of the latter. Much of the evidence for this theory first emerged from work in developmental psychology. Keil and collaborators carried out a series of experiments with children (kindergartners, second graders, and fourth graders) to investigate whether their concepts

---

[8] Later, in Section 2.4, I will argue that it is a mistake to view the theory theory as an alternative theory of conceptual structure and to view it as a direct rival to the prototype theory.

can be identified with lists of features (as in prototype theory), or with theories (as in the theory theory). In these experiments, the typical setup consists in reading a short story to the children and then asking them questions concerning categorization. Experimenters sometimes repeat the story, ask the children to repeat the entire story, and encourage children to deliberate at some length and justify their categorization decisions. I will outline two well-known examples that illustrate the types of experimental protocol used. One story used by Keil (1989, 162) in his experiments described animals living on a farm who neigh, eat oats and hay, and are saddled and ridden by people. The animals are examined by scientists, who find that they have the insides of cows and the blood and bones of cows. Moreover, their parents and offspring are cows. Then children were asked what they thought the animals were, horses or cows. They were encouraged to justify their categorizations in conversation with the experimenters. Another story described taking a raccoon, shaving some of its fur, dyeing it black with a white stripe down its back, then inserting a sac of smelly odor into its body. Again, children were asked what they thought the resultant animal was, a raccoon or skunk. In the first case, most younger children said the animals were horses, while most older children said they were cows, and in the second case, most younger children said it was a skunk, while older children said it was a raccoon (Keil 1989, 164–182). The results of these experiments suggested to Keil and colleagues that for many concepts children exhibit a shift from relying on superficial features to relying on explanatory theories. This shift occurs as early as the preschool years for some concepts and as late as fourth grade for others. After the shift, and into adulthood, thinkers use causal theories in performing categorization tasks rather than relying on characteristic or typical features (Keil 1986; 1989).

Evidence for the theory is not restricted to developmental studies. Other empirical work demonstrates that concept learning in adults occurs more readily for concepts that have features that are correlated by causal connections rather than ones that are not correlated in this way. For example, adults show a strong tendency to cluster features among which a causal link can readily be made rather than ones that do not exhibit this tendency. In learning hypothetical disease categories in a concept-learning experiment, participants found it easier to link dizziness to earaches and weight gain to high blood pressure, rather than dizziness to weight gain and earaches to high blood pressure (Murphy & Medin 1985, 302). Other evidence comes from work on conceptual combination: Combining nouns denoting concepts to form a compound concept often requires thinkers to draw on background beliefs rather than simply select overlapping features. For example, an *expert repair* is a repair done by an expert, whereas an *engine repair* is (probably) not a repair done by an engine but a repair done to an engine (Murphy & Medin 1985, 306). We understand such conceptual combinations because we have theories about experts, engines, and what is involved in carrying out repairs.[9] Finally, there is considerable evidence showing that in feature-listing experiments, the features that people choose to list in connection with a concept vary widely with context in conformity with people's background theories. As mentioned in Section 2.3, different features are associated with the concept PIANO in different contexts. Similarly, when it comes to the concept NEWSPAPER, if one specifies the context of building a fire, the feature "flammable" may be listed, but not in other contexts (Barsalou 1993). To generate such features, people draw on broader theoretical information. Thus, the theory theory considers concepts to be embedded in theories and implicated in informational structures with considerable causal content, and this account is at loggerheads with the one provided by prototype theory, which conceives of concepts as clusters of features.

In this section, we have encountered two currently dominant theories of concepts in cognitive psychology, and in the previous section, we outlined two competing theories of concepts in cognitive neuroscience. There might appear to be a natural alliance between these two opposing theories and the two theories we encountered in the previous section. In particular, prototype theory seems to be compatible with modal or sensorimotor accounts of concepts, since the features associated with many prototype concepts are often posited to be perceptual representations.[10] If the features are not modal, some theorists speculate that these features are themselves concepts that can in turn be represented as sets of features, and that this process will eventually terminate in purely sensorimotor features. Of course, the claim that all concepts can ultimately be decomposed into percepts is a long-standing tenet of empiricism, which is why some philosophers have referred to this research program in cognitive science as "empiricism" (Weiskopf 2007) or "neo-empiricism" (Machery 2007;

---

[9] There is a large body of work on conceptual combination that I will not be able to summarize or address here, but it should be mentioned that some of it purports to show that the prototype theory accounts for the phenomena better than the theory theory (see e.g. Hampton 1997; Hampton 2006).

[10] See Barsalou (2016, 1133) on this point: "Compressed representations, such as CCRs [cross-modal conjunctive representations], are essentially the same kind of representations as prototypes in cognitive theories of concepts ... According to prototype theory, statistically likely features are extracted from category exemplars and conjoined in a prototype that represents the category conceptually. Notably, prototypes are not amodal symbols arbitrarily linked to exemplars. Instead, the features of exemplars appear in the prototype that covers exemplars, following various possible forms of data compression, as for CCRs."

2.3   Empirical Accounts: Cognitive Psychology

McCaffrey & Machery 2012). If this program could be carried out, it might provide a way of linking the prototype theory to modal theories of concepts. The empiricist program to construct concepts out of percepts is widely thought to have failed, as mentioned earlier in this section, but even if it could be made to succeed, much more would have to be done to show how the non-perceptual aspects of prototype theory (e.g. feature weights) could be implemented neurally without introducing amodal elements and undermining the main claim of modal theories. Meanwhile, the theory theory seems consistent with accounts of concepts that emphasize their context-dependence and flexible structure, since researchers who consider concepts to be embedded in theories tend to think that different portions of those theories may be active in different contexts, as seen above. But relating these theories more directly is a daunting task, since it is one thing to say that both theories emphasize flexibility and context-dependence but quite another to show how one might map onto the other. Moreover, both theories face the same challenge of somehow discerning some fixity amidst the instability. Hence, whatever resonance exists between modal theories and prototype theories (on the one hand) and flexible theories and theory theories (on the other) is merely suggestive and the connections between them are, at least for the time being, somewhat tenuous.

In providing an account of the structure of concepts, these theories also effectively supply individuation or identity conditions for concepts, or ways of distinguishing one concept from another, or of saying what makes something the concept that it is. For example, on a prototype theory of concepts, what makes something a concept of APPLE is that it consists in representations of the features associated with apples, each weighted based on its importance or centrality to the concept. Moreover, what it is for the concept to be accessed or deployed on a given occasion is for those representations to be jointly active (along with their accompanying weights) in some way. Similarly, on the theory theory, the concept APPLE is often identified with the theory in which it inheres. That is what makes it the concept that it is, and the concept is accessed when that theory (or some significant part of it) is activated. These brief hints already suggest that both theories face considerable challenges in providing individuation or identity conditions for concepts. In particular, identifying concepts across individuals or even within individuals across times is not a trivial task in either case. For example, on the prototype theory, one would have to say which or how many of the features have to be held in common, and how similar the weightings would have to be, for different structures to be representations of the same concept. The difficulties are at least as formidable on the theory view: Does every change in theory lead to a change in concepts, and if not, how much change or difference can one tolerate? But setting aside these difficulties for the time being, I want to argue that these theories may not be rival theories of concepts at all, but might be tracking different kinds.

The debate between the prototype theory and the theory theory has led to something of a standoff in cognitive psychology. In response to the impasse, various philosophers and cognitive scientists have argued that the prototype theory and theory theory are not genuine rivals. Following Weiskopf (2009, 168), we can divide these theoretical proposals into two groups: pluralist theories and hybrid theories. The basic difference is that pluralist theories posit *different kinds* of conceptual representation, while hybrid theories consider that there is just *one kind* of conceptual representation incorporating distinct components that include different types of information. Hybrid theories typically claim that the component of a conceptual representation accessed on any given occasion depends on the task and that each concept consists of a single complex entity consisting of two or more different representations in different representational formats. Sometimes hybrid theories are supposed to include rule-based and similarity-based representations rather than theory-based and prototype-based representations (e.g. Close, Hahn, Hodgetts, et al. 2010). Pluralist theories, on the other hand, claim that different theories about concepts in cognitive science are actually discussing different kinds of entities, and that each of our concepts may be associated with two or more different representational types. I will now put forward two related considerations for supporting a pluralist account, one based on the methods used by the two theories and the other based on their respective explananda. In earlier work (Khalidi 1995), I argued that the prototype theory and the theory theory rest on different bodies of evidence involving disparate experimental methods. When it comes to the prototype theory, much of the experimental evidence derives from tasks that require rapid categorization of stimuli or snap judgments made under time constraints without a surrounding context. By contrast, results that support the theory theory mainly derive from experimental setups that involve explaining and justifying classifications or inferences in the context of a broader cognitive exercise (e.g. listening to a narrative) and they do not usually include time constraints or measures of reaction times. Accordingly, I hypothesized that the prototype theory and theory theory were pitched at different "levels of explanation" (bearing in mind the caveats from Chapter 1 against conceiving of levels hierarchically). With the benefit of over a quarter century of additional

2.3 Empirical Accounts: Cognitive Psychology

empirical work, this hypothesis still seems plausible, since the results that are taken to support the two theories remain largely disjoint and the types of behavioral and verbal measures are of a different order. Neither theory has been displaced by the other and there has been no real convergence between them. A second source of support for the claim that the prototype theory and theory theory are not genuine rivals derives from the fact that they are focused on different explananda, namely concept *activation* and concept *possession*, respectively. Even though the prototype theory has the resources to provide an account of concept possession, as I briefly indicated above, the methods used are supposed to gauge activation on a particular occasion. By contrast, the theory theory is primarily focused on assessing concept possession. Much of this empirical work compares children with adults or children at different developmental stages. The main objective is not to assess conceptual activation at a particular moment, but to issue a verdict about concept possession by certain individuals or groups based on the totality of evidence gathered. Hence, given that the methods used are largely disjoint and the explananda are distinct, I would maintain that these theories are not addressing the same phenomena.

It may be objected here that the theories are more plausibly interpreted to be addressing different explananda regarding the *same kinds*, rather than different kinds altogether. Just as we might have a theory that explains what it is for a creature to possess a heart and another theory that explains what a heart does, the theory theory explains what it is to possess a concept and the prototype theory explains what it is to activate a concept. But the analogy is not apt, since the two theories of concepts do not even agree on how to individuate their subject matter. It is not as though prototype theory accepts the theory theory's account of concept individuation and possession, and then proceeds to tell us how those very entities are activated on different occasions. Moreover, the causal processes being investigated in each case tend to be somewhat different. The prototype theory is best at explaining categorization and recognition over short time scales when words or images are shown to participants in the absence of a broader theoretical or practical context. The theory theory tends to be better at accounting for reflective or deliberative categorization and inferential judgments that require integrating information from various sources, often involving longitudinal studies.[11] This adds some credence to the hypothesis that the

---

[11] These differences might also be related to "dual-systems" or "dual-process" models of cognition, which posit two cognitive systems, a fast, implicit, and automatic system and a slower, explicit, and more deliberate system. Prototype theory accords with the type of rapid thought processes

prototype theory and theory theory are not rival theories of concepts, but are tracking different kinds. The causal processes that they engage in may interact or intersect in various ways, as I will try to indicate in the next two sections, but they remain somewhat distinct. Moreover, I will argue in the following section that some versions of the theory theory individuate concepts partly on the basis of etiology, or with reference to their causal antecedents, and not just with reference to synchronic causal powers. This lends further support to the conclusion that they are investigating different cognitive kinds. This means that, strictly speaking, there are no such things as concepts simpliciter; for clarity, it may be best to give at least one type of entity a different label.

2.4 A Functional Account of Concepts

Philosophical discussions of concepts (or word meanings) have been dominated in the past several decades by the divide between internalism (or individualism) and externalism, which was outlined in Section 2.1. As already mentioned, externalists hold that concepts are individuated by determinants that are external to the mind of the agent, while internalists maintain that the subject's internal perspective is what individuates a concept. When it comes to the concept APPLE, roughly speaking, the externalist says that what makes it the concept that it is are the thinker's causal connections to the apples in the thinker's environment, whereas the internalist holds that the concept is individuated by its intrinsic character (e.g. the elements of the definition, or the features in the prototype, or the tenets in the theory). One problem with the way these positions are often characterized is that it is not sufficiently emphasized that what matters to the externalist is not so much what is currently in the thinker's vicinity but rather what was causally efficacious at the time of concept acquisition. The various thought experiments proposed to motivate the externalist position usually identify the concept with an external determinant in the context of acquisition, not the context in which the concept is accessed or activated, even though this is not always clearly articulated. Thus, externalists effectively individuate concepts *etiologically*, based on their ontogenetic causal history.

associated with the fast and automatic cognitive system (System 1), whereas responses associated with the theory theory are more in keeping with the slower and deliberate cognitive system (System 2). On some accounts, these two cognitive systems operate independently and can issue in different responses to certain cognitive challenges such as categorization or inference (see e.g. Evans & Stanovich 2013). Moreover, these systems are often regarded as belonging to different kinds or types of cognitive system, not just different token systems (Samuels 2009b).

2.4 A Functional Account of Concepts

While many cognitive psychologists are avowedly internalists about concepts, most philosophers profess externalism.[12] The reason that psychologists (and neuroscientists) tend not to embrace externalism is not difficult to ascertain. Externalism distinguishes among concepts (e.g. APPLE, ORANGE, etc.) based on their causal origin or history, whereas most scientific disciplines and subdisciplines seem to categorize phenomena on the basis of their causal powers or efficacy. More to the point, if one's aim is to explain behavior, there would appear to be a commonality to behaviors exhibited by individuals who share internal or "narrow" states. This commonality is of interest to psychology when it comes to understanding, say, the discriminatory abilities of individuals or the distinctions they make, without regard to the underlying reality or social context (see e.g. Block 1986). It has therefore been difficult for philosophers to convince some psychologists to take externalism seriously when it comes to the individuation of concepts, at least explicitly (for a seminal exchange illustrating the depth of the divide, see Rey 1983; Smith, Medin, & Rey 1984; Rey 1985). Notwithstanding the fact that most philosophers are externalists, some have been swayed by reflecting on scientific taxonomy or psychological explanation to say that internalism or individualism is the only sound taxonomic strategy in science (e.g. Fodor 1987). But this may be because they have been misled by reflecting on some scientific disciplines to the exclusion of others. It is true that in large swathes of the natural sciences, what matters for taxonomic purposes is not the diachronic features of the phenomena being investigated but rather their synchronic properties, specifically their causal powers. Chemists do not usually distinguish among molecules of glucose based on whether they have been artificially synthesized in the lab or are the result of photosynthesis in plants. However, in many other sciences, including some of the physical sciences, some taxonomic categories do track causal origin, trajectory, or history. I have argued elsewhere (Khalidi 2021) that etiological kinds are widespread in science, notably in such sciences as cosmology, geology, and biology, for a number of bona fide scientific reasons (cf. Burge 1986, 18–19). If we grant that there is nothing in principle to prevent scientific inquiries from individuating phenomena based on etiology, does this apply to cognitive

---

[12] Hampton (2006, 84) writes: "Although many philosophers ... have identified major difficulties with descriptivism, preferring to fix conceptual contents in terms of extensions (an Externalist theory of concept individuation), the large majority of cognitive psychologists still subscribe to this basic descriptivist position." According to the PhilPapers 2009 Survey of Philosophers: 51.1 percent favor externalism about mental content, 20.0 percent internalism, and 28.9 percent other.

science in particular? And if so, why should a scientific inquiry into concepts need to individuate them based on their causal origin or history?

Over the past few decades, philosophers have made a strong case for the claim that folk psychology, at least in some circumstances and for certain purposes, individuates concepts based (at least partly) on etiology. But that does not mean that cognitive science does. Is there any evidence that scientific research programs do so, and moreover, that they have good reason for doing so?[13] Some work in developmental psychology mentioned in the previous section, in which psychologists attempt to ascertain whether or not children possess concepts of certain animals or artifacts, are arguably continuous with our folk psychological practices of concept attribution. The same goes for many other research programs in cognitive psychology, social psychology, and educational psychology. In many such domains, there is a concerted effort to determine which concepts individual thinkers possess or have mastered in various experimental conditions. When developmental psychologists conclude that a kindergartner has acquired the concepts ALIVE and DEAD (Bascandziev, Tardiff, Zaitchik, et al. 2018; cf. Carey 1985), or a preschool child possesses the concepts ANIMAL and ARTIFACT (Greif, Kemler Nelson, Keil, et al. 2006), or a two-year-old has the concept CAUSE (Waismeyer, Meltzoff, & Gopnik 2015), or a three-month-old infant has the concepts CAUSE, COST, and GOAL (Liu, Brooks, & Spelke 2019), they are doing so, in large part, on the basis of certain synchronic causal abilities that they have, that is to say, their responses, discriminations, preferences, and related behavioral and cognitive capacities. But they are also basing themselves, at least partly, on etiology. That this is the case can be seen by looking a little more closely at some of the examples just cited.

When a three-month-old infant is ascribed the concepts CAUSE, COST, and GOAL, this is done on the basis of rather minimal discriminatory capacities, having to do with looking times at certain experimental stimuli. By comparing the durations of looking times at different scenes showing goal-directed behavior, experimenters conclude that infants at such an early age possess the concepts in question. Briefly, infants generally exhibit longer looking times at inefficient compared to efficient (or cost-effective) goal-directed causal behavior on the part of other human agents, indicating that they are surprised by such inefficient behavior. This in turn, signals to the researchers that they are able to make the distinction

---

[13] Burge has been perhaps the most vocal exponent of the view that a science of psychology is and ought to be externalist (or anti-individualist, to use his term). This has been a consistent theme in his work from Burge (1986) to Burge (2010), but he focuses mainly on perception and perceptual states.

between behaviors that are cost-effective and those that are costly. Yet, the researchers admit that their experiments and those of other developmental psychologists do not reveal "how richly ... infants represent the costs and goals of other people's actions" (Liu, Brooks, & Spelke 2019, 5). In other words, the infants are able to make some of the relevant discriminations though they might not possess complex representations. Work on children at twenty-four months shows that infants are able to make causal inferences, which implies that they are not only distinguishing instances of causation from non-instances, but deploying the concept CAUSE to infer a cause–effect relationship between two events, indicating further progress along the conceptual trajectory (Waismeyer, Meltzoff, & Gopnik 2015).[14] These two-year-olds still do not have a sophisticated understanding of causation but based on their experiments, the researchers claim that they can distinguish causation from correlation in intervening on the world. In a similar vein, Greif, Kemler Nelson, Keil, et al. (2006) conclude that preschool children make a distinction between animals and artifacts and have the corresponding concepts (viz. ANIMAL, ARTIFACT), based on the fact that they ask different types of questions when confronted with unfamiliar exemplars from each category. In their experiment, three- to five-year-olds posed more questions and made more guesses about functions and behaviors for artifacts than animals, whereas they made more category guesses and asked more questions about niche or location for animals than artifacts. These researchers ascribe the concepts of ANIMAL and ARTIFACT to preschool children while acknowledging that they "may not be able to verbalize the abstract differences between causal patterns associated with living kinds and with artifacts" (Greif, Kemler Nelson, Keil, et al. 2006, 459). Finally, a body of research in developmental psychology shows that while preschoolers have an undifferentiated concept ANIMATE/ACTIVE/REAL, which maps onto the word "alive," older children undergo a process of conceptual change, after which they acquire the concept ALIVE. In addition, Bascandziev, Tardiff, Zaitchik, et al. (2018) demonstrate that some six-year-olds can be induced to acquire the concepts ALIVE and DEAD with training. Even though a thorough mastery of the concepts may not occur ordinarily until much later, they show how certain kinds of training can result in acquiring these concepts for some six-year-olds. In all three cases discussed, while psychologists ascribe concepts to children based partly

---

[14] This work does not make explicit mention of the concept CAUSE, though it is clear from the experimental results that the children are engaging in causal inference, which (at least) involves possessing the concepts CAUSE and EFFECT.

on their causal powers of discrimination, recognition, categorization, and inference, at the same time, the concepts ascribed go beyond their bare abilities when they are "narrowly" conceived. One might well ask why the experimenters attribute richer concepts to the children than may seem warranted by their narrow responses and behavior.

There would appear to be at least two reasons that cognitive psychologists judge that children possess concepts like CAUSE, COST, GOAL, ANIMAL, ARTIFACT, ALIVE, and DEAD, rather than a suite of more rudimentary concepts that reflect their (presumably) more impoverished understanding, recognitional capacities, discriminatory abilities, and so on. The first reason is that children and adults inhabit a single world and children are in contact with the same external stimuli as adults. Possession of a concept marks a certain cognitive achievement and indicates that the thinker in question is able to successfully interact with and navigate some aspect of the world, and psychological theories aim to explain these abilities. Hence, concepts can be seen to be anchored in their origins in our shared world. It is true that psychologists also aim to understand the differences in the ways that different human thinkers conceive of the world (children and adults, members of different cultures, speakers of different languages), but these differences can be described against a background of shared concepts. Indeed, as many philosophers have pointed out, these differences can *only* be described if we presuppose a base of shared concepts (e.g. Davidson 1974). Another reason is that the children are on a developmental trajectory that will, in the overwhelming majority of cases, result in their becoming competent users of language and attaining the full-blown adult versions of these concepts. Rather than ascribe different concepts at every developmental stage, and attribute an entirely different conceptual repertoire at each stage, psychologists regularly say that the children possess these concepts, yet they have an incomplete understanding of them. It is always possible to say something like: The child has the concept of ANIMAL, she just does not know very much about animals. Since these children are in the process of mastering these concepts and becoming full-blown members of our linguistic community, we use the resources of natural language to capture their thoughts, and this practice involves using the concepts lexicalized in the language of their community. Thus, despite rare explicit assent to externalism by cognitive scientists, some research programs in cognitive psychology appear committed to individuating concepts not just on internalist or individualist grounds, but also externalistically, and they have good reasons for doing so. The rationale for this resides partly in the fact that these thinkers inhabit the same environment from which these

2.4 A Functional Account of Concepts 61

concepts derive and to which they apply, as well as the fact that they are or will become part of the same linguistic community.

If concepts in cognitive science are ascribed to agents based partly on synchronic causal powers and partly on etiology, both factors playing a role in determining whether an individual thinker possesses the relevant concept, that also indicates a certain continuity with folk psychological practices of concept attribution. Even though externalists tend to emphasize causal origin as fixing the content of concepts and thoughts, our ordinary folk psychology does not ignore internalist features altogether. According to a pure or strict externalist account of concepts, what makes my concept APPLE that very concept is its etiology, in other words, my causal history with actual apples and with other language users. On a standard externalist account, if I have acquired the word "apple" by way of direct or indirect contact with apples and with members of my linguistic community, I thereby possess the concept APPLE and my word "apple" means APPLE. But if I am under severe misapprehensions about apples and think that they are brown, starchy, and grow underground, and cannot discriminate them from potatoes, then it would be unusual in most everyday circumstances to ascribe to me the concept APPLE. In ordinary conceptual ascriptions, we do indeed give considerable weight to etiology when it comes to the individuation of concepts, but not to the exclusion of synchronic factors, namely the behavioral and cognitive causal powers of individual thinkers. Folk psychology gives some weight to the synchronic abilities of agents, particularly their recognitional, discriminatory, and inferential capacities. Sometimes externalist and internalist accounts are thought to deliver two different concepts of concept, call them concept$_E$ and concept$_I$, or two notions of conceptual content, wide and narrow content. But there is another way to conceive of the situation: Concepts are possessed by individuals in virtue of etiology *and* causal power, and concepts are individuated by both factors in tandem. This goes as much for folk psychology as for many areas of cognitive psychology, as I have tried to argue. Admittedly, I have used just a few examples from the voluminous empirical literature to corroborate this claim, but the methods used and the reasoning deployed are very widespread. Moreover, I have deliberately chosen relatively recent examples from influential research groups to indicate that this is representative of current practices of concept attribution in cognitive psychology, which builds on a body of work on concepts and conceptual development undertaken over the past several decades.

Having argued that much work in cognitive psychology can be understood to individuate concepts both internalistically and externalistically,

I want to propose some convergence between this approach and at least some versions of the theory theory of concepts, which was discussed in the previous section. If the theory theory is understood not as a theory of conceptual structure, but rather as a theory of concept possession, then the account of concepts that is implicit in the research programs cited above can be reconciled with it. One main claim of the theory theory is, as Carey puts it, that concepts "must be identified by the role they play in theories" (1985, 198) and "conceptual role at least partly determines the content of concepts" (2009, 502). This functional or causal-role approach to individuating concepts and determining concept possession can be reconciled with externalism, at least if functions are identified not just narrowly but widely.[15] That is, when ascertaining whether a thinker possesses a certain concept, one attends not just to their categorizations narrowly construed but also to the relevant environmental causes and etiology. By contrast, proponents of the prototype theory, who are more interested in questions of concept activation, tend to endorse an internalist account of concepts (e.g. Hampton 2006). Since etiological individuation need not coincide with individuation according to intrinsic causal powers, this further supports the claim that the two approaches are identifying different cognitive kinds, as I argued in the previous section. Moreover, as I have already suggested, these two theories can be seen to be pitched at different levels of explanation, as these levels were characterized in Chapter 1: relatively "closed systems," each of which is causally integrated and somewhat autonomous from other systems. One way to characterize the difference between the two theories is in terms of what Marr (1982, 22–31) labels the "algorithmic" and "computational" levels, respectively.[16] Marr's theoretical framework can be used to provide an additional reason for considering prototype theory to be pitched at a different level from theory theory. That is because it is natural to think of prototype theory as providing an algorithm for concept activation. As mentioned in the previous section, prototype theory provides a procedure for activating a concept once the weighted values of the associated features have reached a certain threshold.

---

[15] To be more precise, I am arguing for a hybrid (narrow-wide) construal of functions. A hybrid theory of functions of this kind has been articulated by Griffiths (1993), who applies it to the case of biological functions as well as human artifactual functions.

[16] In much earlier work (Khalidi 1995), I hypothesized that the "levels" in question correspond to what Dennett has called the "design stance" (prototype theory) and "intentional stance" (theory theory). I now think that Marr's framework provides a better basis for understanding the differences. Kitcher (1998) has suggested that there is a natural convergence between Marr's algorithmic level and Dennett's design stance. She also identifies a convergence between Dennett and Marr

2.4 A Functional Account of Concepts

By contrast, the theory theory of concepts is more closely related to Marr's "computational level." At the computational level of analysis and explanation, the emphasis is on what concepts enable cognitive agents to achieve and why they possess the concepts that they do. For instance, infants who acquire the concepts CAUSE, COST, and GOAL are able to make certain discriminations and discover certain things that they would not be able to do without that concept. As the researchers put it: "An early-emerging sensitivity to the causal powers of agents, when they engage in costly, goal-directed actions, may provide one important foundation for the rich causal and social learning that characterizes our species" (Liu, Brooks, & Spelke 2019, 17747). Moreover, infants possess these concepts, at least in part, because they inhabit a world of causal agents who are attempting to achieve their goals while incurring low costs, or because their ancestors did so and were naturally selected to think in these terms. A number of philosophers have interpreted Marr's computational level as having an externalist dimension (see e.g. Burge 1986; Kitcher 1998; Egan 2014; Shagrir & Bechtel 2017). In particular, Kitcher (1998, 14) claims that Marr's project shows that science need not be methodologically solipsist, since his computational level "makes essential reference to factors beyond the subject's skin in characterizing psychological states." Similarly, Shagrir and Bechtel (2017, 209) write that the "why aspect" of the computational level "forces researchers to look to the structures in the world that the organism engages through its visual system." Not only are environmental and etiological factors causally relevant in explaining cognitive processes at this level, it also bears repeating that they enter into the *individuation* of cognitive kinds. If we are interested in reliability and success in navigating and interacting with the world, then we will need to individuate concepts partly in terms of their environmental causes and etiology. Moreover, at the computational level, the aim is not to give a structural account of concepts (as the prototype theory does), but to give a functional account, in terms of what possessing a concept enables a thinker to do.

I have been arguing that concepts are cognitive kinds individuated in terms both of synchronic causal powers and etiology. Concepts endow their possessors with certain causal powers or cognitive abilities, though it would be hopeless to try to assign a specific set of canonical abilities

---

when it comes to the implementational level and physical stance, respectively. However, Kitcher (1998, 14–15) proposes that Dennett's "intentional stance" is a fourth level, distinct from Marr's "computational level," and she criticizes Dennett for holding that the intentional stance cannot be rendered scientific, since Marr's program shows that it can be.

to each concept. Moreover, each concept operates only in conjunction with other concepts, so there is no requisite set of abilities attached to each concept in isolation from others. In addition, I have argued that concepts are also identified with their contextual and historical determinants and individuated partly in terms of those determinants. How are these two determinants of concepts, synchronic and diachronic, related to each other? Rather than think of these two factors as distinct components of conceptual content, they can be conceived as joint determinants of conceptual content, combining to provide individuating conditions for concepts. This gives us a way of transcending a dilemma faced by "two-factor" or "dual-factor" theories of concepts (e.g. Block 1987; Carey 2009), at least if these factors are understood as distinct components of concepts.[17] Since the internalist factor and the externalist factor tend to pull in opposite directions whenever a thinker exhibits significant ignorance or harbors a misconception, that is, whenever one's thoughts are seriously out of step with the world or the linguistic community, it is not clear what dual-factor theories would say about a thinker's concept in such cases. Is it the narrow concept or the wide concept that takes precedence? Rather than thinking of concepts as being resolvable into two possibly opposing factors or components, it is more in keeping with both folk and scientific taxonomy to regard concepts as amalgamating both factors. But if concepts are individuated both synchronically and diachronically, do we have specific criteria for how much disparity in causal powers to tolerate, or how much leeway to give those whose concepts have the right etiology, in individuating concepts or determining which concepts are possessed by a thinker? There seem to be no hard-and-fast criteria, whether in folk or scientific psychology, but that should not undermine the entire enterprise of conceptual taxonomy. In many cases in science, such as in assigning species to higher phylogenetic taxa, we weigh diachronic and synchronic factors in making a determination about classification, without having sure-fire rules for doing so. Here too, we balance different considerations and issue in a verdict as to whether the thinker has the concept in question or not. As in other cases in scientific taxonomy, there may be borderline cases in which there is no fact of the matter as to whether a thinker possesses a concept or not. Sometimes

---

[17] Block sometimes talks in terms of narrow and wide "determinants" of meaning rather than aspects or kinds of meaning (see e.g. Block 1986, 620), but it is not clear how he views the relationship between them. The account I am proposing may be closer to the "long-armed" conceptual role semantics of Harman (1982).

2.4 A Functional Account of Concepts

the etiology is right but the synchronic causal powers are lacking; at other times, the agent is making the appropriate discriminations and categorizations but the external antecedents are not the ones we expect.[18]

Before concluding this functional account of concepts, it is necessary to say something about the role of language in individuating concepts. In finding the right words to capture a thinker's thoughts, we are not simply forcing a fluid mental reality to conform to a rigid representational medium. The process of allocating the correct linguistic labels to a thinker's thoughts is not like trying to measure temperature using a crude and inaccurate thermometer, because language is a representational medium that allows us to capture subtle differences among thinkers whose concepts might not seem to be in perfect alignment. Earlier, I said that cognitive psychologists decide to attribute rich adult-like concepts to infants and children rather than ascribing a suite of more impoverished concepts. Developmental psychologists choose not to report the children's beliefs in terms of some alternative set of concepts, CAUSE*, GOAL*, ANIMAL*, ARTIFACT*, and so on, but instead rely on the familiar set of adult concepts. That is not a distortion of their mental lives because language enables us to pinpoint areas of disagreement using a common stock of concepts. So when describing the mental lives of preschool children, instead of saying that they have some concept ANIMAL* rather than ANIMAL, we can say that they have the concept ANIMAL but they do not know that animals are multicellular, or that they have a common lineage, or that they breathe oxygen, or whatever other discrepancies we might find between our views and theirs. The same goes for any two thinkers who harbor different conceptions, make distinct associations, or hold disparate beliefs in connection with any given concept. In some cases, we do resort to introducing novel concepts and coining new terms when faced with thinkers whose behavior and responses cannot be captured at all in concepts expressed in natural language, but these are the rare exception, even when it comes to infants and other atypical thinkers. (One such case was mentioned above: Carey attributes an undifferentiated ANIMATE/ACTIVE/REAL concept to preschoolers.) Language provides a way of fixing the individuation conditions of concepts despite variability in the way they are deployed across contexts

---

[18] I do not have space here to address the notorious cases that have figured so prominently in the philosophical literature, such as the WATER or ARTHRITIS cases, which are used to motivate externalism. But in at least some of these cases, there may be no determinate answer to the question of concept possession. Recall that on this hybrid account, it is not enough simply to be causally related to a certain environment or community, one also needs to have certain powers of discrimination, inference, and so on.

or by different individuals. By giving a role to the words of shared, public, natural languages in the individuation of concepts, one incorporates the *social* etiology of concepts as well as their *natural* etiology or their causal origin in a shared world.

Some cognitive scientists claim that words are what give the *illusion* of conceptual fixity or relative stability (e.g. Casasanto & Lupyan 2015), but this is misguided not only because we can accurately capture subtle individual differences using language as just indicated, but also because language can be seen to have a role in the development of concepts. As Lupyan and Thompson-Schill (2012, 20) put it: "Words may matter far more for conceptual representations than previously considered, in that some concepts may only attain sufficient 'coherence' when activated by verbal means." Language can be seen to play an active role in the evolution of conceptual thought, both ontogenetically and (more speculatively) phylogenetically. When it comes to ontogeny, a large body of research shows that words support individuation, inductive inference, and causal reasoning. For example, providing twelve-month-old infants with a word highlights commonalities among objects that go undetected in the absence of a word (see Waxman & Gelman 2010 and references therein). This same research demonstrates that the word-concept link is not just one of pure association and that words do not merely play the role of an attentional spotlight. In fact, "the conceptual status of words comes not from the sound of a word itself, but rather from its role within the linguistic and social system in which it is embedded" (Waxman & Gelman 2010, 107). As for phylogeny, the evidence is more speculative, but there is also reason to think that language played a prominent role in the development of full-fledged human conceptual thought. Many philosophers have theorized about the indispensability of language for augmenting cognition and enabling certain conceptual achievements that would not have been possible without language. Clark (1998, 173–174) writes:

> The role of public language and text in human cognition is not limited to the preservation and communication of ideas. Instead, these external resources make available concepts, strategies, and learning trajectories which are simply not available to individual, un-augmented brains.

On such a view, the evolution of language and the evolution of the human conceptual repertoire are intertwined in such a way that it is difficult to prise them apart. Our concepts coevolved with the words of natural languages, and language provides part of the scaffolding that enables human conceptual thought. If linguistic symbols play an active role in shaping mental phenomena, both ontogenetically and phylogenetically, then it

2.4 A Functional Account of Concepts

is misleading to think of language as providing the *illusion* of stability to concepts. Rather, linguistic symbols are instrumental in determining conceptual identity. This provides another reason for casting doubt on the notion that language is an inadequate representational medium that distorts the rich content of thought.

To sum up, I have argued that in both folk and (at least some areas of) scientific psychology, concepts are individuated with regard to both synchronic causal powers and diachronic etiology. This method of individuating concepts pertains primarily to investigations of concept possession (as opposed to activation), which are pitched at what Marr designated the computational level of explanation. Moreover, this approach is most closely aligned with the theory theory of concepts, by contrast with the prototype theory, which is pitched at the algorithmic level. Since these different theories pertain to different explanatory levels, I proposed that they investigate different scientific domains, which are populated by different kinds. Rather than a unitary cognitive kind, *concept*, these theories are attempting to understand different constructs, call them *concept$_1$* and *concept$_2$*. A question that arises here is how to relate the hypothesis that there may be different kinds associated with the label "concept" in these different domains to recent evidence from neuroscience, particularly the theories surveyed in Section 2.2. I argued in Section 2.3 that despite the fact that there might seem to be a close alliance between the theories proposed on the basis of neural evidence and those based primarily on behavioral measures, the connections are likely more tenuous. In fact, the neural evidence pertains to what Marr dubs the "implementational level." The modal theory of concepts is clearly framed in implementational terms since it posits that the very same neural networks that process perceptual information also process concepts and that they do so using the resources of the sensorimotor systems in the brain. Rather than expect a direct correspondence between this neural or implementational theory and theories framed in algorithmic (or computational) terms, modal theories of concepts can be seen to operate with different taxonomic categories. What they are calling "concepts" might not correspond either to the construct identified at the algorithmic level or that investigated at the computational level, but some third construct, *concept$_3$*. Hence, I am proposing that the theories of concepts under discussion in this chapter are investigating (at least) three different kinds that are implicated in distinct causal processes.[19] This may seem like a profligate agenda to multiply concepts beyond necessity, but the

---

[19] Machery (2005; 2009) also argues that there is a plurality of concepts of CONCEPT in contemporary cognitive science, but he takes this as grounds for eliminativism about concepts and for denying

alternative is to suppose that there is a neat correspondence among the entities tracked by these different research programs with their different methodologies and their investigations into disparate causal processes, some of which are individuated with reference to context and history while others are not, and some of which aim at explaining reflective thought and inference while others are primarily interested in recognition and spontaneous judgment. It would be more surprising to discover that they were all really theorizing about the very same kind of thing.

2.5 Objections and Replies

In the previous section, I tried to outline an account of concepts that considers them to be real kinds in cognitive science, in accordance with the account of cognitive kinds that I elaborated in Chapter 1. I can anticipate a number of objections that the account may elicit, so I will try in this section to raise and respond to the most prominent ones. Since many of the replies are already implicit in the previous discussion, I will address them rather briefly and without detailed elaboration.

One objection would draw attention to the similarities between this theory of concepts and the theory theory of concepts. Given that the theory theory is prone to some well-known objections, it is natural to ask whether this account is open to the same objections. Two of the most powerful objections to the theory theory are the circularity objection and the holism objection. The circularity objection says that the theory theory equates concepts with theories or parts of theories, but theories are themselves constituted at least in part by concepts. This leads to a "mereological paradox" (Laurence & Margolis 1999, 44). The response to this objection is to say that concepts should not be *equated* with theories or parts of theories. Concepts are inextricable from theories and are not independent of them, but that does not mean that they are identical to miniature theories or sets of theoretical tenets. If concepts are individuated in part narrowly, based on a thinker's ability to perform certain cognitive tasks associated

---

that *concept* is a natural kind. I will not try to respond directly to his arguments, but my main disagreement with Machery is that I think that there is a single kind that plays a dominant role in higher cognitive processes such as those described in this section, while he thinks that three different alleged kinds play this role (in addition to prototypes and theories, he discusses exemplars). Machery (2009, 242–243) thinks that each of these categories corresponds to a kind, but withholds the term "concept" from any one of them. By contrast, I have argued that there is a continuity between our pre-theoretic notion and the category that features in some scientific research programs, which justifies reserving the term for that category.

2.5 Objections and Replies

with those concepts, then a possessor of a concept holds certain theoretical tenets associated with it, whether explicitly or implicitly. But that does not make the concept equivalent to those theoretical tenets. This account of concepts is a functional rather than a structural one; part of the mistake is to conceive of the theory theory as being a structural rival to the prototype theory. The holism objection is related; it says that if a concept is associated with a theory or part of a theory, then any change in theory results in a change in the concept, with the result that concepts differ constantly within and across individuals. I would respond to this worry by briefly expanding on what I said in the previous section regarding concept individuation. Cognitive scientists (as well as the folk) usually represent a thinker's mental life using the resources of natural language. The terms of natural language align the thinker's concepts with those of the community and any differences between them can be represented against a background of common concepts. Hence, concepts do not differ incrementally with every difference in belief, and concepts do not change slightly with every change in belief. Differences and variations in thought can be captured by indicating disagreement among beliefs using a common stock of concepts.

Another objection to this theory of concepts might question its affinity to a rather different account, namely the interpretivist account associated with Dennett's "intentional stance" (e.g. Dennett 1987). That account of concepts is widely perceived to make concept ascription subjective or response-dependent, which many cognitive scientists doubt is compatible with a true science of the mind. In previous work (Khalidi 1995; Khalidi 1998), I favored an interpretivist account of concepts, which is indeed response-dependent. But even though the interpretivist account contains important insights (including the inextricability of concept and theory, or meaning and belief), I now think that response-dependence does not accord with scientific or folk taxonomic practice. I have argued that concept possession is partly a matter of a thinker's synchronic causal abilities (sorting, recognizing, discriminating, categorizing, inferring, and so on) and partly a matter of the thinker's relations to external determinants (including social determinants) in the history of the thinker. These determinants of concept individuation do not depend ultimately on how others assess them, but on facts about thinkers themselves and their relations to the world.

A different objection to a broadly functional account of concepts would question whether one can in fact equate the content of each concept with its functional role. The reply to this concern is that this is not meant to be a reductive account. The individuation of concepts is based on what individuals do with those concepts, taking etiology and environment into consideration. But there is no attempt to identify a requisite set of abilities corresponding to each concept. There is no fixed set of hoops that a thinker must jump through to demonstrate that he or she possesses the concept APPLE or ORANGE (or for that matter, CAUSE, ANIMAL, or LIFE). There may be different ways of exhibiting possession of a concept, and anyway, most concepts are acquired in close confederation with other concepts. It is not a coincidence that many studies in cognitive psychology examine a small number of closely related concepts together (e.g. CAUSE, GOAL, and COST; CAUSE and EFFECT; ANIMAL and ARTIFACT; ALIVE and DEAD). Hence, there is no question of isolating each concept and specifying its functional role in isolation from other concepts. This does not mean that cognitive scientists do not have ways of gauging whether individual thinkers possess certain concepts as opposed to others, as already suggested in looking at some of the empirical work on children's concepts. But as is clear from considering this work, concepts should be thought as abilities or *capacities* rather than *objects* or concrete particulars. In Chapter 1, I said I would indicate, where applicable, how cognitive kinds fit into certain overarching superordinate kinds. In this case, I am calling for revising a common ontological view among some philosophers and cognitive scientists that takes concepts to be objects instead of capacities.[20] As I have already argued, these cognitive capacities should be individuated functionally, where functions are understood in hybrid terms, both causally and etiologically.

It may also be objected that the account of concepts that I have outlined is too abstract to be suitable for cognitive science. In particular, it does not seem to square with the usual cognitive scientific claim that concepts are to be identified with *mental representations*. I think the right response to this objection is to say that concepts can be abstracta and mental representations at once. They are abstract kinds, but they are also representational in nature, since they are individuated by what they represent. This representational content is supplied jointly by their causal history and their causal role, though I have not tried to supply a full-blown theory of mental representation. However, concepts should not be thought of as concrete entities or objects with a definite location or spatial dimensions. That is what we would expect to find at the implementational level of explanation, not at the computational level. In some discussions in cognitive science "mental representation" is used synonymously with "neural representation," but the question of the existence and nature of neural representations is an

---

[20] This view is perhaps most clearly instantiated by Fodor (1998, 2–4), who regards it as a fundamental error on the part of cognitive science to think of concepts as abilities rather than objects.

independent one, and moreover, I have given reasons for doubting that there will be a neat correspondence between the computational and implementational levels.[21]

There is a related objection that would cast doubt on the claim that the construct "concept" identified in these different research programs really picks out different kinds. It may seem both taxonomically extravagant and prima facie improbable that completely different kinds have (unwittingly) been identified by these different research programs.[22] To this objection, I would reply that there may well be important relations between kinds identified at the implementational, algorithmic, and computational levels (though I cannot be sure that real kinds have already been identified in this vicinity at the first two levels). But we should not expect a relation of reduction or a one-to-one correspondence between them. To make this more plausible, consider the concept ALIVE, as discussed at the computational level. As we have seen, current research in cognitive psychology concludes that this concept is connected with such concepts as DEAD, is typically acquired by children by ages ten to twelve (in North America), can be induced via training to younger children, enables children to answer questions and make distinctions concerning animals and plants, and is associated with a "vitalist" theory in biology (Bascandziev, Tarfdiff, Zaitchik, et al. 2018). Meanwhile, other researchers in psychology have noted that children and adults respond distinctively to stimuli that move in certain ways, reacting to them as though they are alive (see e.g. Mandler 1992). This automatic response to (apparently) living things is clearly not tantamount to judging that something is truly alive nor that the concept ALIVE or LIFE correctly applies to it. It would seem to pertain to a system or mechanism that takes as input certain perceptual stimuli and issues a fast and non-deliberative response, perhaps something at the algorithmic level. Moreover, the relationship between the output of that system and the considered categorization judgment is likely to be highly indirect. We often react immediately to a stimulus (a branch shaking in the wind, a piece of fluff propelled by a draft) as though it were alive, but go on to suppress our initial reaction. This response may be one input into the conceptual system but that does not mean that it is a component of

---

[21] For a recent account of neural representation, see Shea (2018).
[22] While advocating a pluralist view of concepts, Weiskopf (2009) makes the case that the different kinds are subordinate kinds belonging to a single superordinate kind, *concept*. I do not have space to counter his arguments for this conclusion directly, but I think the considerations that I have brought forth argue for the existence of entirely different kinds. See Machery (2009, 243–245) for some replies to Weiskopf's arguments.

the concept itself. In the specific context of the representation of animacy, Mandler proposes a number of different levels from the perceptual to the conceptual and hypothesizes that "human infants represent information from an early age at more than one level of description" (1992, 602). There is no reason to conclude that the representations at each level simply correspond to or are combinations of those at lower levels.

A final objection would accuse this account of concepts of being maximalist or intellectualist rather than minimalist, to revert back to a distinction from Section 2.1. Given the tight link that I have posited between concepts and language, this theory of concepts would seem to apply exclusively to language-using creatures, or at least those creatures who are on track to acquire language. While some philosophers would be sympathetic to such a view, many others would not, claiming that there are no good grounds for denying concepts and conceptual thought to, say, nonhuman animals (for detailed discussion, see Camp 2009). In the previous section, I argued that concepts are ontogenetically and phylogenetically associated with language. Language is not just a representational medium that happens to individuate concepts, it is coeval with them. Therefore, even though non-linguistic thinkers can possess concepts, it is doubtful that their concepts can be individuated as finely as those of language-using creatures, and there will be far more inherent vagueness in the concepts that they possess. This does not mean that non-linguistic thinkers cannot have concepts, just that there will be more cases in which it is genuinely indeterminate which concepts they possess.

2.6 Conclusion

The causal-etiological account of concepts that I have outlined in this chapter suggests that concepts are a real kind in our cognitive ontology. To briefly recap the main conclusions of this chapter, it may be worth reframing them in terms of the central research questions that were outlined in Section 2.1. When it comes to the first question about the identity of concepts, I have argued that in the cognitive or computational domain, concepts should be considered abstract functional entities rather than concrete particulars like neural structures or processes. Since they have intentional content, they can be thought of as mental representations as long as they are not considered concrete representational entities. As for the individuation of concepts, their representational contents derive from their wide functions and are a combination of etiological and causal factors. This means that concept acquisition is not just a matter of the thinker having the right causal history but also

2.6 Conclusion

being able to make the right distinctions and inferences – though there is no set of requisite abilities or achievements associated with each concept. This, in turn, implies an account of concept possession according to which having a concept is a matter of a thinker's inferential and discriminatory capacities, as well as their etiology. Finally, this view of concepts does not address the question of concept activation, which I have argued is more properly addressed at the implementational level.

Another way of framing this account of concepts is in terms of some of the central theoretical divides mentioned in Section 2.1. I have defended a hybrid account of conceptual content that combines internalist and externalist elements. I have also argued against the modal (or sensorimotor) theory of concepts, at least as a theory of concepts at the computational level, rather than at the implementational or algorithmic levels. This account has a strong affinity with the theory theory of concepts, though I have also argued that there may be a different type of psychological entity at the algorithmic level that may conform to the prototype theory. Like the theory theory, the account is holistic about conceptual content. Some standard objections against holism have been briefly addressed. I have argued against a response-dependent account of concept identification and individuation, on the grounds that concept possession is determined by a thinker's abilities and history, not ultimately by the responses of other members of their communities. Finally, the account lies between minimalism and maximalism, in that language may not be strictly necessary for the possession of concepts, but it helps determine the fixity of concepts as well as their acquisition and possession.[23]

It may be thought that this chapter has focused on kinds of *individual* concepts, not on the kind *concept* as such. Never mind the cognitive kinds APPLE, ORANGE, CAUSE, ANIMAL, and LIFE, what about the cognitive kind *concept* itself? What are its identity conditions and how does one individuate it? An analogy can be drawn here with the kind *species* and its subkinds, *Panthera tigris* (tiger), *Drosophila melanogaster* (fruit fly), and *Acer palmatum* (Japanese maple). Note that in both cases, these subkinds do not bear a straightforward relationship of subordinate kinds to a superordinate kind. In the case of species, the higher taxa, like genera and families are true superordinate kinds. By contrast, the kind *species* and the kind *concept*, bear a relation to their subkinds (particular species and particular

---

[23] Camp (2009) distinguishes three grades of conceptual thought: minimalist, moderate, and intellectualist. The moderate conception seems to be the most congenial to the account that I have defended in this chapter.

concepts, respectively) that is more akin to the determinate–determinable relationship, like the relation of *color* to *violet*, or the relation of *length* to *one millimeter*. Thus, if we grant that specific concepts (e.g. APPLE, ORANGE) are causal-etiological cognitive kinds, then there is no further question as to whether *concept* itself is a real kind in the cognitive domain. If the determinates are real kinds so is the determinable. Still, if specific concepts are identified with their functional roles, what are concepts *in general*? I have argued that cognitive scientists have ways of individuating concepts such as CAUSE and ANIMAL and means of assessing whether a thinker possesses them, but do they have ways of discovering whether a thinker has concepts at all, as opposed to not having concepts? The synchronic and diachronic conditions for possessing concepts are closely related to the conditions for having beliefs or thoughts. I will not attempt to identify the conditions for thought or cognition or general, which are notoriously difficult to spell out. But there is some consensus that cognition involves an ability to represent the world in a stimulus-independent way (see e.g. Camp 2009; Beck 2018). For full-fledged *conceptual* cognition, mere stimulus independence may not be enough. Camp (2009, 303–304) argues that conceptual thought combines stimulus-independent representational ability with the capacity to recombine representations, and that this "makes a practical difference for achieving the most basic aim of thought: using information about the world to solve problems and facilitate one's survival and flourishing." Hence, conceptual thought can be distinguished from thought in general by this ability to represent the world in a stimulus-independent way and recombine those representations in such a way as to interact flexibly with the world. This characterization also combines synchronic and diachronic factors, since it involves the ability to navigate the world flexibly as well as the ability to form representations that derive from the world. It also confirms the importance of language for concept possession, since language is a medium that facilitates representation and keeps track of representational recombination, even though language may not be strictly necessary for having concepts. As I argued at the end of the previous section, there is nothing to preclude all thinkers, including nonlinguistic creatures, from having concepts. But in the case of nonlinguistic thinkers, it will often be difficult to determine whether they possess concepts, and there may be no determinate answer to the question of which concepts they possess.

CHAPTER 3

Innateness[1]

Every human brain is born not as a blank tablet (a tabula rasa) waiting to be filled in by experience but as 'an exposed negative waiting to be slipped into developer fluid'.
– E. O. Wilson (quoted by Tom Wolfe)

Nature ... is what we are put into this world to rise above.
– Katharine Hepburn in *The African Queen*

3.1 Introduction

Innateness is somewhat unusual as a candidate for a cognitive kind since it may be more naturally considered a property than a kind. However, as I argued in Chapter 1, the distinction between properties and kinds is not a principled one and does not run very deep. Though kinds generally consist of clusters of properties, kinds can themselves sometimes be conceived of as properties. For example, the kind *Panthera tigris* is characterized by a number of etiological and causal properties, including the property of being a carnivore, but *carnivore* can also be considered a kind with individual members, such as members of the species *Panthera tigris*. The same would seem to go for *innateness*. It can be considered a property of particular cognitive capacities, but it can also be considered a kind, which includes particular cognitive capacities among its instances. In this discussion of innateness, I will focus primarily on cognitive capacities because I think that these are the most likely candidates for innate cognitive entities. In this context, cognitive capacities can include such things as the mindreading capacity in humans and birdsong in some species of birds. Therefore, in what follows I will focus on these kinds of cognitive capacities in attempting to justify the claim that innateness is a real cognitive kind.

---

[1] This chapter is an updated and modified version of Khalidi (2016a).

To say that innateness is a real kind is not to imply that any particular cognitive capacities or kinds of capacity are in fact innate. But it does mean that there is a meaningful question to be asked about whether a given cognitive capacity is innate. It may be that the extent of the innate human cognitive endowment is considerably less extensive than traditional nativist philosophers believed or many contemporary cognitive scientists posit. Nevertheless, I will argue that there is a non-vacuous kind, whose members can be distinguished from nonmembers. Another feature of the innateness category is that it appears to admit of gradations. It seems at least coherent to posit that some cognitive capacities are more innate than others. Unlike some other cognitive kinds, whether something belongs to the kind *innateness* might seem to be a matter of degree, as when we say that the capacity for birdsong in one species of bird is more innate than in another (though they both possess *some* degree of innateness). This may seem to rule it out as a kind-category, since kinds might be thought to be distinguished by an "unfathomable chasm" rather than a "mere ordinary ditch," to use Mill's expressions (1843/1882, 152). But I will try to show how these differences of degree can be understood on the model of innateness that I will propose, and will also try to justify the claim that this does not jeopardize the kindhood claim.

Though it originated as a folk or vernacular category, innateness has featured prominently in contemporary controversies in cognitive science.[2] There are ongoing debates concerning whether the linguistic faculty is innate to the human species, the extent to which numerical, spatial, and causal cognition are innate, and the degree of innateness of moral rules and religious concepts, among others. Yet, in addition to these first-order debates about the innateness of some of our basic cognitive capacities, there is a second-order debate about the viability of the innateness concept in the first place. Some scientists and philosophers have voiced reservations about the very category of innateness, questioning its suitability for rigorous scientific theorizing and doubting that it corresponds to a real cognitive kind. If they are right, then that would mean that the debates about the nature and extent of our innate cognitive endowment are pointless, since they revolve around a discredited concept. In response, I will attempt in this chapter to defend the concept of innateness against its critics and to establish that it is indeed a real kind in the cognitive domain.

---

[2] The OED defines innate, as applied to "qualities, principles, etc. (esp. mental)," simply as "opposed to acquired." The earliest usage cited is from Thomas Hoccleve, *The Regiment of Princes* (c. 1420), who speaks of "innat[e] sapience [i.e. intelligence]."

3.1 Introduction

To appreciate the objections to the concept of innateness, it is necessary first to be acquainted with some recent efforts to define or characterize the concept of innateness. What is it for a cognitive capacity to be innate? Various attempts have been made to provide an explication of innateness that accords with contemporary scientific theorizing. Recent accounts of innateness include the following:

- *Canalization:* Innateness is canalization, where a phenotypic endstate is canalized to the degree to which the development of that endstate is insensitive to a range of environmental conditions under which the endstate emerges (Ariew 1999; 2007; Collins 2005).
- *Entrenchment:* Innate traits are ones that are generatively entrenched, in the sense that they appear early in development and have a number of later developing traits dependent on them (the degree of entrenchment is correlated with the number of traits depending on them) (Wimsatt 1999).
- *Psychological primitiveness:* Innate traits are explanatorily primitive in the domain of psychology, that is, the proper explanation of their acquisition lies outside of the domain of psychology (Cowie 1999; Samuels 2002).
- *Triggering:* An innate cognitive capacity is one that has a disposition or tendency to be triggered on the basis of an environmental input that is informationally impoverished by comparison to the resultant cognitive capacity (Stich 1975; Khalidi 2002; Khalidi 2007).
- *Closed process invariance:* An innate trait is one that develops across a range of normal environments and the proximal cause of its development is by a closed process or processes, where a closed process is one that tends to produce one or very few outcomes (and where closure is a matter of degree) (Mallon & Weinberg 2006; Weinberg & Mallon 2008).

Some of these proposals attempt to account for uses of innateness in a wide array of sciences, including microbiology, genetics, evolutionary biology, and so on, while others are intended to be more restricted to cognitive science. In what follows, I will be concerned with trying to rehabilitate a concept of innateness that applies primarily to the cognitive sciences. Rather than trying to characterize innate traits or phenotypic endstates in general, I will attempt to explain what it is for a cognitive capacity to be innate. As will become clear in due course, the characterization applies primarily to representational or information-bearing cognitive states. So it is quite possible that the innateness category has outlived its usefulness in many areas of biology, yet continues to be of value in cognitive science.

3.2 Critiques of the Innateness Category

All the attempts to elucidate innateness mentioned in the previous section have been deemed beside the point by other theorists on two principal grounds. The first criticism of the innateness category has to do with its alleged association with discredited essentialist and typological ideas in biology, while the second criticism claims that the category of innateness runs together a number of unrelated features or properties. Though these critiques are often combined, I think that they are distinct and will only focus here on the second criticism.[3]

The second criticism of the category of *innateness* charges that, rather than being a unified category, *innateness* is a multivalent category, combining a number of disparate criteria (Griffiths 2002; Bateson & Mameli 2007; Mameli 2008; Mameli & Bateson 2011; Shea 2012a). Closely related to this criticism is the contention that the different factors that influence people's judgments about innateness are not always correlated and that they can therefore lead to unwarranted inferences from one of those features to one or more of the others. The objection here is not that the features themselves are problematic but that the features, some of which are scientifically respectable taken individually, are not always correlated. The natural response to this objection consists in pointing out that innateness may be a polythetic category or a cluster category, which combines several of the features posited by the analyses mentioned and that these features are imperfectly correlated. By contrast with a monothetic category, the features are not singly necessary and jointly sufficient for the application of the category. Many other scientific categories, especially biological ones,

---

[3] I will not try to respond here to the first criticism, which has been made in a number of papers (Griffiths, Machery & Linquist 2009; see also Griffiths & Machery 2008, Griffiths & Stotz 2008, Linquist, Machery, Griffiths & Stotz 2011). But it is worth pointing out that even if the vernacular concept of innateness has some of the associations that these critics have identified, that may not render it unfit for scientific theorizing. Many scientific concepts originate as folk concepts before being refined and revised in order to make them suitable for scientific theorizing. In a recent empirical study of the innate concept among laypersons and scientists, Knobe and Samuels (2013) argue that members of both groups can engage in a process of "filtering" tainted concepts, dissociating them from unwanted prescientific associations. Hence, even when vernacular concepts have been implicated in prescientific or discredited scientific theories, scientists (and the folk) are capable of jettisoning their problematic features, especially when thinking explicitly and making considered judgments (as opposed to making snap decisions under time constraints).

are of this kind (e.g. biological species). Since scientific categories often appear in exception-prone empirical generalizations, there is usually a risk of making an unsuccessful inference from one of the features associated with a scientific category to another.

But the claim has also been made that the innateness category consists of a "clutter" rather than a cluster of criteria (Bateson & Mameli 2007; Mameli 2008; Mameli & Bateson 2011). Though these critics do not give a precise way of distinguishing a clutter from a cluster, they have likened the category of innateness to the pre-theoretical category JADE. As is now well known, jade comprises two distinct minerals, jadeite and nephrite, with entirely different chemical compositions. These substances share some of the same macro-properties (e.g. color, hardness) by sheer coincidence, but there is no superordinate kind *jade* that unites them. To escape the fate of jade, these critics state that a cluster theorist "has to give an account of the ... properties that constitute the cluster, of the causal processes that connect such properties and cause them to tend to co-occur" (Mameli & Bateson 2011, 441). These researchers argue that some of the properties associated with innateness are scientifically sound but they are not always reliably correlated or causally linked. In what follows, I will try to respond to the challenge posed by the critics of innateness and will both outline the main properties associated with innateness and give a preliminary account of some of the causal connections among these properties. These properties are conceptually distinct and the causal connections among them may not be obvious, but I will maintain that there is considerable empirical evidence to suggest that they are reliably causally linked, enough to provide grounds for thinking that innateness constitutes a cluster category rather than a clutter.[4]

3.3 Innateness as a Cluster Category

A cluster category, with non-strict causal relations among the properties included in the cluster, is just what we would expect in a theoretical category in the biological and cognitive sciences. The question is, what are the associated properties in the case of *innateness*, and how are they related? We have already encountered some of them, and others can be readily gleaned from recent research in cognitive science. What follows is a tentative list

---

[4] A similar proposal has been made by Samuels (2007), but the features he associates with innateness are somewhat different from those I posit, and he continues to consider psychological primitiveness to be the primary feature associated with innateness, as I will go on to argue.

of properties or features associated with innate cognitive capacities, which may not be exhaustive:

- *Triggering* (or more properly, *triggerability*)*:* can be acquired in conditions of relative informational impoverishment
- *Lack of learning:* need not be acquired as a result of processes such as inference, conditioning, association, exploration, experimentation, repeated observation, and imitation
- *Early onset:* is acquired relatively early in ontogeny
- *Invariance:* is acquired across a broad range of environments
- *Canalization:* is buffered against environmental variation
- *Pan-cultural:* is present in all cultures, even though it may not be universal or monomorphic
- *Informational encapsulation:* is insulated from other cognitive content, functions independently of other cognitive systems
- *Cognitive impenetrability:* resists modification by other cognitive capacities
- *Critical period:* is acquired only or most effectively within a developmental window

In this section, I will try to show that, at least according to some prominent research programs in cognitive science, there are robust causal links among these properties. Cognitive scientists regularly make inferences and provide explanations that posit causal relations that situate these properties in a web or network, with some of these properties being causally prior to other properties in the network. Although I will not offer a comprehensive account of this causal network, I will try to provide enough examples to make a plausible case for the existence of such causal connections. Many of these examples will be drawn from a prominent nativist research program in developmental psychology, the "Core Cognition" research program championed by Carey, Spelke, and others. This is obviously not the only research program in cognitive science to use the category of innateness, but it is one of the most prominent ones to do so and at least some of its results concerning the development of human cognition have been widely accepted. It is therefore convenient as a relatively established scientific paradigm that has made widespread use of the category of innateness.

Triggering and Lack of Learning: If a cognitive capacity can be acquired on the basis of a trigger, that is to say that it can be acquired as a result of an impoverished input. While it may not make sense to say that an input is impoverished in an absolute sense, it is possible to rule that an input is impoverished relative to the resulting output, the output being the cognitive state or capacity that is acquired as a result of that input (Khalidi 2002; 2007). One of the most widely used argumentative strategies in cognitive science, the argument from the "poverty of the stimulus," relies on just such considerations to reach the conclusion that some particular cognitive state is innate (Laurence & Margolis 2001). But if something is disposed to be acquired as a result of an impoverished input, then it is plausible to say that it need not be acquired as a result of a learning process on the part of the agent, including trial-and-error, extensive observation, imitation, inference, or "any process that treats information derived from the world as evidence" (Carey 2009, 453). The reason is that treating the world as evidence involves an inferential process wherein information is processed and this generally involves something more than just minimal input from the environment. To be sure, the difference between learning something on the basis of information derived from the environment and having something triggered by an environmental input may be a matter of degree. Yet, learning and triggering can be thought to represent two ends of a continuum. To the extent that a cognitive capacity is disposed to be triggered, it can be acquired without learning. That is why some cognitive scientists consider that when a cognitive capacity has been acquired on the basis of an impoverished input, that is evidence that that capacity has not been learned (and hence that it has a substantive innate component). As Carey (2009, 196; emphasis added) puts it in discussing infants' capacity to represent agency: "In some cases the age of the infants, *along with considerations of limitations on the inputs they could possibly have experienced*, casts doubt on some plausible learning accounts." If one thinks of a trigger as an input that is impoverished relative to the resulting output, then it will lead to the acquisition of a cognitive capacity without the need for learning. Many research programs in the cognitive sciences treat innateness and learning as complementary notions, and though some also countenance degrees of innateness, it is safe to say that there is a negative correlation between degree of innateness and amount of learning.

Lack of Learning and Early Onset: If a cognitive capacity does not need to be learned then it can be expected to be acquired relatively early in development. If the acquisition of a cognitive capacity does not involve a process of deriving evidence from the world, then there is less of an obstacle to acquiring that capacity early in development. Developmental psychologists often argue from the early onset of certain cognitive capacities to the conclusion that they have not been learned, at least given other circumstantial evidence. For example, Carey (2009) holds that one of our core cognitive capacities includes a capacity to recognize others as agents and to represent others' mental states. Moreover, she thinks that this capacity of "core cognition" is not learned partly on the grounds of early onset and the lack of sufficient perceptual input. As Carey (2009, 196) puts it in a passage already quoted above: "In some cases the age of the infants ... casts doubt on some plausible learning accounts." Even though the causal link from innateness to early onset may be weak, inference in the other direction is stronger, since it is unlikely that a cognitive capacity could be acquired so early in development by a process of learning. In other words, there is a causal link between lack of learning and early acquisition and there is also an inference to the best explanation from early acquisition to lack of learning. This causal link is taken for granted in so much recent psychological research in the past two decades that it explains the explosion of experimental work on infants younger than six months old, as well as the development of elaborate techniques for appraising the mental states of such preverbal and unreliable participants (most famously, the "violation of expectation" experimental paradigm).

Triggering and Invariance: Even though there are crucial exceptions, there is a natural link between triggering and invariance. If a capacity is capable of being triggered, manifesting itself on the basis of a relatively impoverished input, then it is likely to emerge in a variety of different circumstances and will therefore be relatively invariant across a range of environments (cf. Khalidi 2007, 109). This causal relationship can be illustrated by various types of studies, but it is perhaps most clearly revealed in research on animal cognition in which animals are reared in a variety of environments. For obvious ethical reasons, these types of experiments, which involve raising infant animals in a range of different environmental settings, are conducted primarily on nonhumans. The idea behind these experimental manipulations is to determine whether a cognitive capacity is relatively invariant across environments. If a capacity emerges in all or a wide range of different environmental contexts, this is regularly taken to show that it does not require specific experiences with the environment, and hence that it is merely triggered by the environment. There has been considerable research into spatial orientation in a variety of animal species, specifically regarding whether they use geometric or nongeometric information to orient themselves in unfamiliar environments, or to re-orient themselves having been disoriented in familiar environments (see e.g. Hermer & Spelke 1996). In nonhuman animals, this aspect of spatial cognition is often investigated by raising animals in environments with a variety of different spatial configurations to ascertain whether this leads to important differences in their abilities to spatially orient themselves. Consider recent research into spatial orientation in chicks, which concludes that the cognitive capacity to reorient using geometric information is largely innate. Chiandetti and Vallortigara (2008, 144) observe: "No differences between chicks reared in circular-, rectangular-, or c-shaped cages were apparent in the ability to reorient using purely geometric information (i.e. in the absence of any featural cues)." Partly on this basis, they conclude that "the results reported here for chicks ... suggest that animals encode geometric information in the absence of (or with minimal) experience of surfaces of different lengths connected together at right angles" (Chiandetti & Vallortigara 2008, 144). That is, they infer that the cognitive capacity is triggered based on invariance across a range of training environments, since there is a causal link between a capacity's being capable of being triggered and its emerging across a wide range of environments. Again, the *inferential* link between invariance and triggering rests on a *causal* link in the opposite direction.

Invariance and Canalization: A cognitive capacity that is invariant is not always canalized, but one prominent way of achieving invariance in a cognitive capacity is by means of the ontogenetic device of canalization. A cognitive capacity that is canalized is buffered against environmental variation and develops according to a relatively fixed developmental pathway or a small set of such pathways. It is "programmed" to proceed along a finite number of different "channels" in such a way that precludes intermediate or hybrid developmental outcomes. Invariance in a biological setting can be achieved by means of a process of canalization, as in the acquisition of birdsong in many species of birds. In bird species in which the development of song is highly canalized, being buffered against environmental variation, the outcome (acquiring the species-specific adult song) will be invariant across a range of environments. Invariance can be achieved de facto in other bird species, where acquisition is not canalized, but it is so achieved only if a wide range of environments will contain the input needed in order to lead reliably to acquisition. It is worth observing that canalization does not lead to invariance in the sense of constancy of outcome, much less constancy of outcome across all possible environments. Rather, canalization produces a relatively *limited number* of outcomes across a very *wide range* of environments (i.e. relative rather than absolute invariance). In this case, as in some of the subsequent causal links to be explored below, invariance is not strictly causally linked to canalization, but given certain background conditions and plausible assumptions about implementation in a biological system, one prominent way of achieving invariance is by means of canalization. Moreover, the *direct* causal link operates in reverse: Canalization of a trait causes that trait to be invariant across a range of environments. Thus, if a capacity is canalized, it follows that it will be relatively invariant. However, one way that natural selection has devised to secure the invariance of a trait is to canalize it. The adaptiveness of invariance when it comes to certain traits has (so to speak) led natural selection to canalize some of those traits. This is why invariance can be said to be both an effect and a cause of canalization. The causal link here is a functional one in the sense that canalization has the "function" of bringing about invariance. As in the case of other causal-functional links, the effect (invariance) raises the probability of the reproduction of the cause (canalization), which in turn produces (another token of) the same effect, in a positive feedback loop.

Invariance and Panculturality: There is reason to expect, and considerable evidence to suggest, that capacities that are relatively invariant across environments will emerge in all human cultures. A pancultural capacity need not be monomorphic in the species, since it is quite compatible with its arising in all cultures that it be polymorphic or indeed that is a rare trait arising in a few select individuals. Still, invariance across a range of environments suggests that it would be compatible with the whole range of human *cultural* environments. Features of our core cognition are regularly claimed to be impervious to cultural differences in this way. Consider, for instance, recent work on concepts of OBJECT and SUBSTANCE, as manifested in human participants in the United States and Taiwan, native speakers of English and Mandarin, respectively. Since Mandarin is a classifier language while English is a count-mass noun language, some research has suggested that speakers of the former will tend to assume that a new word refers to a kind of substance while speakers of the former will tend to assume that it refers to a kind of object. But Li, Dunham, and Carey (2009) claim that though such effects can be observed in linguistic tasks, they do not occur in at least some nonlinguistic tasks. They conclude:

> The distinction between object kind and substance kind is a central piece of our core ontology, integral to our ability to make sense of and navigate a complex and shifting world. As such, it may not be surprising that, far from being highly malleable, it should prove itself quite resistant to linguistic or cultural influence, part of the shared conceptual endowment of our species. (Li, Dunham, & Carey 2009, 518)

Whether or not they are right to draw this conclusion, it is clear that the authors think that their nativist account of ontological categories links invariance to resistance to cultural influence. The fact that this cognitive

3.3 Innateness as a Cluster Category

capacity is regarded as invariant or not malleable is taken to imply that it is pancultural.[5]

Canalization and Cognitive Impenetrability: If a cognitive capacity is canalized then it is buffered against environmental variation, and this will often imply that it is impervious to input from other cognitive systems or that it is cognitively impenetrable. A canalized cognitive system needs to be protected from being modified, which means that its representational content should be resistant to being overwritten by other systems. One way of achieving this is by making the system cognitively impenetrable. Carey (2009, 68) considers one of the main properties of innate "core cognition" to be that it "is never overturned or lost, in contrast to later developing intuitive theories..." Canalization may not always lead to cognitive impenetrability, but unless a canalized system is cognitively impenetrable then it may not be sufficiently buffered against information from the environment that contradicts the information represented in that system. It is plausible that any genuinely canalized cognitive capacity would be shielded from being altered in this way.

Canalization and Informational Encapsulation: Similar considerations suggest that canalization among cognitive capacities may also be achieved by means of informational encapsulation. A cognitive capacity that is informationally encapsulated is one that does not depend on other informational systems to perform its function.[6] If a cognitive capacity is canalized, then it may emerge regularly after other systems and in roughly the same order, but if that capacity is itself one that appears early in development, then it ought not to depend on more developed cognitive systems to perform its function, on pain of being inoperative. Such appears to be the case with the system of spatial orientation in humans, which is dependent only on basic perceptual information to perform its function, as opposed to "higher" cognitive functions. Hermer and Spelke (1996) found that while human adults use both geometric and nongeometric information to reorient themselves in a novel room after being turned around several times to disorient them, children (aged eighteen to twenty-four

---

[5] To say that the distinction between objects and substances is innate is not necessarily to say that the concepts OBJECT and SUBSTANCE are themselves innate. But the innateness of the distinction may facilitate the acquisition of the concepts.

[6] The terms "informational encapsulation" and "cognitive impenetrability" are sometimes used interchangeably, but I am using them here in different senses, based on the way that they appear to be used in the empirical research that I am relying on. It could be said that the sense of informational encapsulation at play here is a rough counterpart, in the cognitive domain, of the notion of "entrenchment" used by Wimsatt (1999), which was mentioned in Section 3.1.

months) rely only on geometric information. Even though this geometric capacity for reorientation is informationally encapsulated, in the sense that it is largely self-contained and independent of other cognitive systems, by the time humans reach adulthood, they are also able to use other, nongeometric information to reorient themselves. This encapsulation of the capacity for spatial orientation in very young children, as well as its task specificity, are taken by Hermer and Spelke (1996) to imply that this capacity is an innate module. As Hermer and Spelke (1996, 227–228) put it: "Both the task specificity and the relative encapsulation of the reorientation process suggest, to a first approximation, that children's reorientation process depends on a 'geometric module …'" Though they have not tested directly whether this geometric module would emerge in a range of environmental settings, Spelke and Newport (1998, 315) argue that this too is plausible, on the grounds that the developmental environment for the subjects in these experiments is different from that of their evolutionary ancestors: "Because the laboratory animals and American children in these studies have not spent their lives in outdoor environments where hills and valleys uniquely specify object positions, but rather in rectangular environments where many symmetries make geometry-based reorientation prone to error, it is likely that this process has been shaped more by evolutionary history than by learning." Researchers infer from the informational encapsulation of this capacity to its modularity, as well as its canalization and lack of learning.

Canalization and Critical Period: Yet another way of bringing about canalization is by means of a critical period.[7] A cognitive capacity that is subject to a critical period is likely to be canalized at least in the following sense. An organism that receives the appropriate input or inputs within the critical period is then sent along a specific developmental pathway, whereas one that does not receive the appropriate inputs fails to be launched on that pathway. The critical period can be thought of as the entrance to a developmental pathway, without which the organism fails to proceed along that pathway or "channel." In some instances, there may even be two or more types of input, which when received during the critical period, determine different developmental pathways. This is one way of understanding

---

[7] A distinction is sometimes made between a *critical period* and a *sensitive period*, the difference being that the former entails a sharp cutoff in the ability to acquire a cognitive capacity while the latter involves a more gradual decline in that ability. However, as numerous researchers have pointed out, there are few if any sharp cutoffs of this sort in cognitive development, so all such periods are more properly thought as sensitive periods. But since it is the more common terminology, I will use "critical period," with the caveat that this does not imply a sharp developmental divide.

3.3 Innateness as a Cluster Category

language acquisition in humans on the principles-and-parameters account (e.g. Baker 2001). Depending on the type of input received from language speakers in their environment during the first few years of life (the critical period), the parameters are set in a certain way and human infants then go on to acquire the syntax of the appropriate language. Canalization occurs because the input received during the critical period directs one along one pathway or another (or in the case in which no input is received, fails to proceed along a pathway at all). This phenomenon is evident in research on the human phonetic repertoire. In many instances, young infants can discriminate a wider range of phonetic contrasts than are made in their native language. If their native language makes a certain distinction, they receive input that observes the phonetic distinction within the critical period and it is consolidated; if it does not, then they receive no such input and their ability to make the distinction is lost. The consolidation of some phonetic discriminations in turn influences the acquisition of yet others, in what has been described as "cascading" critical periods, each constraining and directing the next (Werker & Tees 2005). These successive critical periods can act as channels along which phonetic development takes place. (More will be said about critical periods in Section 3.4.)

These connections among the properties associated with innateness suggest a causal network in which the instantiation of some of these properties leads reliably, though not inevitably, to the instantiation of others (see Figure 3.1). Moreover, in some of these cases, the links are causal-functional, in the sense that the property in question does not just lead in a linear manner to the production of the effect. Rather, it does so precisely because it was selected to do so and in doing so raises the probability that the cause will be reproduced. Hence in some of these cases (indicated in Figure 3.1), there is a causal feedback loop between cause and effect rather than a simple linear relationship.[8] Complex causal interactions among the properties associated with innate cognitive capacities clearly signal a difference with a category such as *jade*, which is given as a paradigmatic example of a "clutter" category by Mameli and Bateson (2011). Unlike the macro-properties of jadeite and nephrite, which are accidentally correlated, the properties of innate cognitive capacities are causally connected in various intricate ways.

There are two ways in which this account of innateness admits of degrees.[9] The first is the one alluded to in the discussion of triggering and learning. These complementary properties are themselves dimensional. To

---

[8] I am grateful to an anonymous referee for Khalidi (2016a) for urging me to clarify this point.
[9] For a more detailed discussion of degrees of innateness, see Khalidi (2007, 102–109).

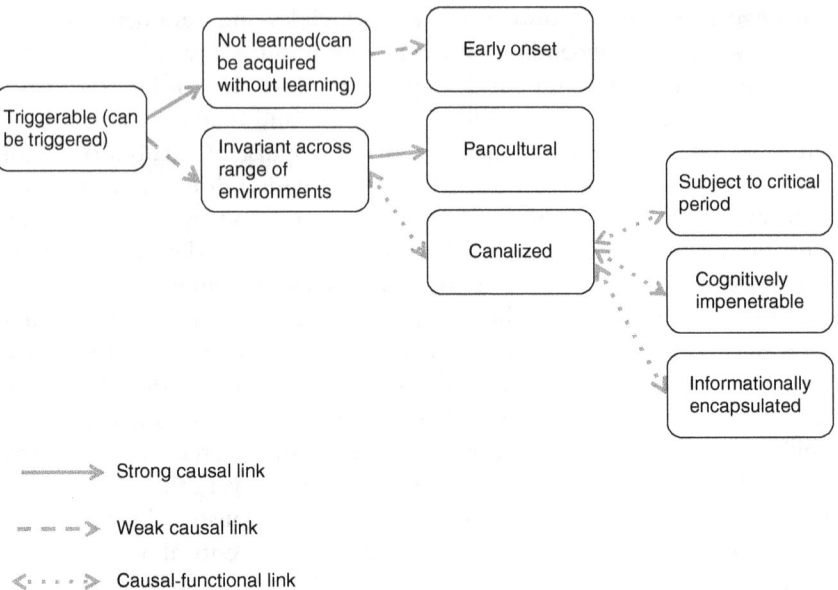

Figure 3.1. Causal network associated with the kind *innate cognitive capacity*

the extent that something is triggered, it is not learned, and this means that a capacity may be more or less innate depending (roughly) on the extent of triggering required to acquire it. In addition, the properties in this cluster are linked by non-strict causal relations. Hence, some instances of innate cognitive capacities may be characterized by more properties in the cluster than others, due to the operation of intervening causes. Those instances that are characterized by more of the properties in the cluster can be said to be more innate than others, even though they are both innate to some degree. This is simply a consequence of the fact that innateness, like many other cognitive kinds, is characterized by a cluster of properties that are linked by causal connections that are not strict. In fact, this is a characteristic of many paradigmatic kinds in the special sciences (and even, I would argue in the basic sciences), which is why such fuzzy kinds are ubiquitous in nature (Khalidi 2013, 65–69).

3.4 Is Innateness a Homeostatic Property Cluster?

Given the characterization of innateness in the previous section, it is natural to regard the category of an innate cognitive capacity as corresponding

3.4 Is Innateness a Homeostatic Property Cluster?

to a homeostatic property cluster (HPC), which conforms to the account of natural kinds developed by Boyd (1989), according to which natural kinds are clusters of properties kept in equilibrium by a certain causal mechanism. A similar proposal has been made by Samuels (2007), who explicitly identifies innateness as a HPC kind. As outlined by Samuels, HPC kinds satisfy three conditions: (i) They are associated with a number of features that tend to be co-instantiated, none of which is necessary for membership in the kind, (ii) a causal mechanism explains the co-instantiation of these features, and (iii) it is the causal mechanism, rather than the associated features, that constitutes the essence of the kind and defines membership in it (Samuels 2007, 23). There is a superficial similarity between my account of innateness as a real cognitive kind and Samuels' account, but I will argue in this section that my proposal differs from Samuels' in three important respects.

The first difference with Samuels' account concerns the particular features that I have proposed are associated with innateness. Though there is some overlap among the features we associate with the kind *innateness*, there is also some crucial divergence. Most significantly, Samuels considers that the central property of innateness, to which other properties are evidentially related, is psychological primitiveness, where a cognitive capacity is psychologically primitive if there is no psychological account of its acquisition. But far from being the central feature of innateness, I would argue that this is not a feature of innate cognitive capacities at all, since it has to do with the type of scientific explanation offered for the acquisition of an innate capacity rather anything about the capacity itself.[10] Moreover, several of the other features that Samuels associates with innateness, such as being *present at birth*, *adaptive*, and *monomorphic*, are ones that I do not consider to be associated with innateness with any regularity. Being present at birth is widely considered in the developmental literature not to be sufficient for innateness, since significant learning is now known to occur in the womb in humans and other organisms (see e.g. Partanen, Kujala, Tervaniemi, et al. 2013). It is clearly also not necessary, since many if not most innate characteristics, in cognition and elsewhere, are manifested well beyond birth, even in late adulthood. Being adaptive is also not strongly correlated with innateness; to think so is to commit a kind of adaptationist fallacy, since many innate cognitive and psychiatric disorders (though perhaps not all) are clearly maladaptive. Finally, monomorphism is no more

---

[10] I have put forward other criticisms of the primitiveness account elsewhere, see Khalidi (2007).

associated with innate features than polymorphism, since in humans as in many other organisms, there are a large number of significant dimorphic or polymorphic innate traits, from innate sexual traits to other phenotypic traits like eye color, hair color, blood type, lactose tolerance, and so on. It is likely that some innate cognitive features may be similarly polymorphic.

A second major difference with Samuels' proposal, is that he does not attempt to reconstruct the actual causal links that obtain between the features associated with the kind *innateness*, contenting himself with discussing "evidential relations" between these features. Although evidential relations are presumably grounded in causal connections, evidential relations can obtain between a cause and one of its effects, between an effect and one of its causes, as well as between two effects of a common cause. To say that features I_1 and I_2 are evidentially linked is to leave open the precise causal relationship between them. Instead, I have tried in the previous section to delineate these causal connections, at least in light of the current state of scientific research. While some of these causal links are straightforwardly linear, others appear to involve feedback loops, as mentioned in the previous section, and involve complex interactions between distinct properties. In order to justify the claim that innateness is a genuine cognitive kind, it is not enough to specify evidential or inferential relations among its associated features. One must show how these features relate ontologically to one another, as I have tried to do by means of causal links. This causal theory of innateness does not conform to the template outlined in Boyd's HPC theory of natural kinds, but it is consistent with the account of real kinds as "nodes in causal networks" that I explicated in Chapter 1 (see also Khalidi 2013).

The third difference with Samuels' account, and perhaps the most important, has to do with the fact that I deny the existence of a causal mechanism that keeps the cluster of properties associated with innateness in homeostasis, as specified by the HPC account of natural kinds. Whereas Samuels regards the "mechanism" as the causal essence of the kind, I think it is problematic to posit a causal mechanism that defines the kind or constitutes its underlying essence, at least if the term "mechanism" is used with anything like its standard meaning. Sometimes, the term "mechanism" is used so loosely that it is roughly synonymous with "cause," but there is a more precise usage that has become very widely accepted, according to which it refers to "entities and activities organized such that they are productive of regular changes from start or set-up to finish or termination conditions" (Machamer, Darden, & Craver 2000, 3). On Boyd's account of natural kinds, there is typically a causal mechanism

3.4 Is Innateness a Homeostatic Property Cluster?

that keeps the properties associated with any given kind in a state of equilibrium or homeostasis, understood roughly along these lines. In the case of innateness, a natural suggestion would be that the mechanism that keeps the properties in the cluster in a state of homeostasis is *genetic*, and that the entities and activities that produce the innate cognitive capacities are genetic in nature. However, there are two related problems with this suggestion. The first is that the genetic cause is likely to be so far upstream from the cognitive capacity that it does not seem possible, in general, to think of it as the single mechanism that keeps the properties in homeostasis. That is simply because there is a looseness of fit between the genetic factors that may contribute causally to the emergence of an innate cognitive capacity and that capacity itself, since there are numerous intermediate processes that may contribute to the outcome. As Spelke and Newport (1998, 291) once observed: "the central accomplishments of recent research in developmental neurobiology are to reveal a host of epigenetic processes through which neural structures develop in accord with a species-typical, intrinsic plan, without either shaping by the environment external to the organism or detailed genetic instructions." The second problem with positing a genetic cause common to all innate cognitive capacities is that it is unlikely that all innate capacities will be underwritten by the same type of genetic substratum. The point is not merely that there is likely to be a different genetic substratum that brings about the innateness of cognitive capacity C_1 and cognitive capacity C_2, but that the genetic substrata in question may not themselves share any important properties in common (beyond being somehow genetically instantiated). This is not to rule out that we may, in the case of some particular cognitive capacity, find that a certain genetic mechanism codes for that capacity. In the most idealized scenario, there may be a specific nucleotide sequence that codes for a protein, which then encourages the formation of synaptic connections between groups of neurons, which in turn are the neural substrate for a putatively innate cognitive capacity, such as the capacity to represent object permanence or analog numerical magnitude. But there may well be other innate capacities for which the causal story is far more complicated, involving multiple regions in the genome, regulatory mechanisms, epigenetic processes, interactions with the environment, and so on. An analogy with innate diseases may help: Consider two diseases that are largely innate, Huntington's chorea and cystic fibrosis, but for quite different reasons and as a result of very diverse causal pathways. Huntington's results from an abnormally long trinucleotide repeat near the tip of chromosome 4, while cystic fibrosis turns out to involve a more heterogeneous collection of mutations (Kitcher 1997, 60). Hence, what it is for a disease to be innate (or partly innate) can be the result of different types of genetic (and epigenetic) processes. The situation is likely to be similar in the case of innate cognitive capacities: What it is for each of them to be innate at the genetic level may differ considerably, so there is little prospect of locating a single type of genetically based mechanism that is common to the development of all innate cognitive capacities.[11] In the absence of a single type of genetic mechanism underwriting all innate cognitive capacities, the regularly co-occurring properties associated with an innate cognitive capacity may be understood not as having a single underlying cause, but rather as being themselves causally interconnected in various ways. That is (as Boyd himself acknowledges at times), there need not be a single mechanism to keep them in homeostasis.[12]

It may be objected here that the mechanism that corresponds to the innateness of a cognitive capacity need not be genetic but may instead be neural. That is, even though the genetic bases of innateness may be diverse, there may be a common neural mechanism or process that corresponds to the innateness of a cognitive capacity. Perhaps there is a certain type of neural profile that all innate cognitive capacities share, such as a certain mode of neural connectivity that renders them capable of achieving a mature state with relatively minimal external stimuli. It is certainly possible that there are neural commonalities corresponding to the cognitive kind *innateness*. But given that I have argued that the central property in the innateness cluster is the disposition to be triggered, it is at least possible that innate cognitive capacities are susceptible to triggering via different routes. The disposition to be triggered by external stimuli is a prime example of a functional property that can be realized differently in different physical structures. To use a simple analogy, the triggering mechanism of a gun is different than the tripwire mechanism that triggers a landmine. In both cases, the result is an explosive process that can be initiated by a relatively weak external stimulus, but this disposition is realized by very

---

[11] Shea (2013) has developed a notion of *inherited representation* that attempts to escape this problem of genetic and epigenetic heterogeneity. Without trying to rehearse this notion in any detail, it may serve to encompass a variety of different genetic and epigenetic mechanisms for encoding cognitive capacities so that they can be passed from one generation to the next. Thus, I would not rule out the possibility of a viable characterization of the mechanism that holds the features listed in homeostasis.

[12] "Either the presence of some of the properties in [a family of properties] F tends (under appropriate conditions) to favor the presence of the others, or there are underlying mechanisms or processes which tend to maintain the presence of the properties in F, or both" (Boyd 1989, 16).

3.4 Is Innateness a Homeostatic Property Cluster?

different structural means. Even in the absence of direct evidence that a disposition to be triggered is multiply realized in the cognitive domain, it seems clear that it is multiply *realizable*. There may be different neural mechanisms leading to the development of a cognitive capacity with minimal triggering or in the absence of learning. Hence, it is at least possible for there to be a one-to-many relation between the cognitive kind *innateness* and its neural correlates. Moreover, it is also possible for the relationship to be many-to-one. Given the relational nature of some of the properties associated with innateness, the very same neural properties that underpin them may correspond to different cognitive kinds in different contexts. This appears to be true of the disposition to be triggered, which is characterized in terms of the relative informational contribution of the environmental stimulus and the resulting cognitive capacity. If triggering is assessed based on the informational content of the stimulus in relation to the resulting cognitive capacity, it is not clear how this could be captured purely in neural terms (even if we had the means to discern the informational content embodied in neural mechanisms). In addition, I have made a similar case elsewhere (Khalidi 2020) for the cognitive kind *critical period*, which is characterized in terms of its place in the lifespan of the organism. The same neural mechanism that correlates with a critical period may be correlated with a different cognitive kind if that mechanism were to operate toward the end of an organism's lifespan. Thus, at least some of the properties in the innateness cluster in the cognitive domain are individuated contextually with reference to environmental or developmental contexts. If in some cases the corresponding neural mechanisms are not so individuated, this would yield a many-to-one mapping between the cognitive kind and its neural correlates. The result is a many-to-many mapping between cognitive and neural kinds.

I have now expanded on three ways in which this account of the cognitive kind *innateness* does not conform to the standard characterization of HPC kinds, as clusters of properties kept in equilibrium by a causal mechanism. Instead, according to the position that I have outlined, *innate cognitive capacity* is a natural kind associated with a cluster of properties related to one another by intricate causal connections, though not held together by a single causal mechanism. This account also allows us to respond to a recent critique of the innateness category that has been put forward by Shea (2012b). One of his principal reasons for rejecting the proposal is that the clustering of properties associated with innateness is especially unreliable when it comes to human beings, mainly on the grounds that inherited representations are not necessarily genetic but can also be epigenetic, cultural and so on. The idea seems to be that some of the properties associated with innate cognitive capacities may be the result not of inherited *genetic* representations, but of (say) inherited *cultural* representations, and this may lead us to conclude falsely that a cognitive capacity is innate when it is not. To use an illustration of my own, suppose that the cognitive capacity to read and write is a human cultural invention that has been transmitted as a result of imitation. If we observe that reading and writing are pancultural and highly canalized, we might be tempted to infer that they are innate. But we would be rash to conclude this, since panculturality and canalization may in fact be the result of an inherited cultural representation and therefore not innate. If that is the worry about innateness, then it seems misplaced. Even if there are other causal properties that tend to generate some of the same properties in the innateness cluster, as I have identified them, that should not lead us to conclude that innateness is not a genuine cognitive kind, though it may make it harder to distinguish innate from non-innate cognitive capacities. There is a rough analogy here with the case of jade, considered earlier, but with the following crucial difference. The fact that the macro-properties of jadeite and nephrite largely coincide for accidental reasons tells against counting *jade* as a natural kind, but it surely does not undermine the case for considering either *jadeite* or *nephrite* separately as natural kinds. Just because many of the properties in the jadeite cluster coincide with those in the nephrite cluster, that should not deter us from thinking that either of them is a natural kind in its own right. Similarly, *innate cognitive capacity* may well be a natural kind even though some of the properties that are causally associated with it are also associated with cognitive capacities that are culturally transmitted and not innate.

In this section, I have tried to show that there are three important differences between this account and Samuels', differences that would rule out considering innateness a natural kind along the lines of Boyd's homeostatic property cluster kinds. Instead, innateness can be considered a natural kind along the lines of the simple causal theory of natural (or real) kinds defended in Chapter 1.

3.5 Objections and Replies

One objection to the argument that I have presented pertains to the very identity of the category that I am purporting to defend. The revamped category of innateness that I have tried to defend might be said to bear little resemblance to the original folk category of innateness, suggesting that we should scrap it and start anew rather than attempt to brush up a tired vernacular category. In this vein, Mameli (2008) compares the concept *innateness* to the concept *mass*, which the special theory of relativity allegedly eliminated and replaced with two concepts, *rest mass* and *relativistic mass*, neither of which coincides with the concept *mass* as found in classical mechanics.[13]

A response to this objection must address the question of conceptual change and continuity, as well as the relationship between scientific concepts and lay concepts (albeit briefly). The descriptive question as to when a concept has been retained and when it has been replaced, and the related prescriptive question as to when we ought to retain a concept and when to replace it, are controversial to say the least. But it seems safe to say that when there is significant continuity between a lay concept and a scientific concept, retention is usually the outcome. Moreover, there are two prominent prescriptive considerations that would tend to favor retention in many cases. First, an entrenched lay concept is sometimes difficult to abandon entirely and attempts to expunge it are often counter-productive and likely to be met with resistance. Second, as long as there is some continuity, it is more conducive to the comprehension and communication of scientific results to express them mainly in terms of existing concepts (wherever possible) rather than introduce altogether new concepts. These points can be illustrated using the concept HEAT as it was used by both scientists and laypersons in the western world until the mid-eighteenth century, that is until Joseph Black's proposal of the theory of latent heat. Until this time, scientists had not clearly differentiated the concepts of KINETIC ENERGY and TEMPERATURE (Carey 2009, 372; cf. Wiser & Carey 1983; Wiser 1988). Whereas kinetic energy is an extensive and additive quantity, temperature is intensive. If two cups of water at the same temperature are added together, the quantity of kinetic energy in the mixture is increased but the temperature of the water remains the same. Once this differentiation was made, all subsequent scientific theorizing on the subject proceeded to distinguish the two concepts. How should we describe this episode in intellectual history or the history of science: Is

---

[13] Mameli's claim concerning the concept MASS is controversial and seems to endorse a view according to which the concept did not survive the theory change from classical to relativistic physics, a view that has been widely disputed. Many scientists and philosophers of science have argued instead that REST MASS should be identified with the classical concept MASS and that the latter concept has not been eliminated at all (see e.g. Earman (1977), Earman & Friedman (1973)). Note also that it may be misguided to insist to an engineer that one should not talk about MASS, but must always distinguish REST MASS from RELATIVISTIC MASS.

it one of elimination or modification? We could say that the folk concept HEAT was abandoned in rigorous scientific thought and replaced by two distinct concepts, KINETIC ENERGY and TEMPERATURE. Alternatively, we could (and often do) say that the scientific concept KINETIC ENERGY is roughly equivalent to the vernacular concept HEAT, and that the latter concept has been modified to differentiate it from the concept TEMPERATURE. In the former case, we have eliminated a folk concept in favor of two scientific concepts, and in the latter case, we have modified a folk concept and rendered it roughly equivalent to a scientific concept. If our linguistic practices are anything to go by, in this case, it would seem that elimination has lost out to modification. The term "heat" in English, like related terms in other languages, survives today and is commonly regarded as a loose synonym for the scientific term "kinetic energy" in certain contexts. If that is the correct way to describe the situation, we can still ask a prescriptive question as to whether this was the right course of action, or whether it might not have been preferable to abandon the folk concept in favor of the two scientific concepts. Though it might still give rise to misconceptions among the folk and lead to occasional mistaken inferences, there are advantages to modifying an existing concept rather than eliminating it altogether. Again, there are two considerations that favor retention. First, since the concept HEAT was highly entrenched, it might not have been feasible to eliminate it altogether. Second, as long as there is some continuity, the sciences are often better served when they can relate their findings to our prescientific theories rather than when they introduce new jargon or specialized language that is not readily accessible to the lay public. We do not normally say that there is no such thing as heat, but teach schoolchildren that heat is different from temperature. Similarly, in the case of innateness, it might be more productive to frame our scientific findings in terms of innateness rather than try to purge it from our vocabulary. Moreover, there is sufficient continuity that it seems feasible here to express our scientific findings in terms of our existing concept. Rather than saying that there is no such thing as innateness, it may be more productive to point out, say, that while it is true that what is innate is not learned, it is not the case that what is innate is always present at birth.[14] This stance

---

[14] Recent detailed investigations of the interactions between lay concepts and scientific concepts indicate that the relationship is more complex than some philosophers have hitherto believed, and may not always involve deference by laypersons to the scientists. For instance, Radick (2012) relates that the late nineteenth- and early twentieth-century geneticist William Bateson resisted the concept HEREDITY but eventually succumbed to widespread usage, indicating that scientists sometimes defer to the lay public.

3.5 Objections and Replies

regarding concept individuation coheres with the account of concepts that I defended in Chapter 2. On that account, concepts are individuated on both externalist and internalist grounds. In this case, the scientific concept and folk concept apply to many of the same capacities (e.g. to reasoning, but not to reading), and there is considerable continuity in the functional-causal profile of the concept (e.g. the contrast with learned capacities, the association with early onset). Hence, there is sufficient reason to consider that the folk and scientific concepts coincide. (Recall also that this does not mean that there is exact coincidence between all the associated beliefs.)

Another, not unrelated, objection suggests itself here. If we attach the label of *innateness* or *innate cognitive capacity* to the entire causal network that I sketched out in Section 3.3 (and represented in Figure 3.1), it may be said that this cluster of causal properties corresponds more closely to the presumptive natural kind *cognitive module*, roughly in the sense first articulated in Fodor (1983), rather than the natural kind *innate cognitive capacity*. In other words, the kind I have described might seem to have a number of distinguishing features that go well beyond the bare kind *innateness*, even *innate cognitive capacity*.

There is likely to be some overlap between the kind *innate cognitive capacity* and the kind *cognitive module* (assuming the latter is also a cognitive kind), and the latter may indeed be a subordinate kind of the former. Though there may be some cognitive modules that are not innate, many of the cognitive modules now posited in cognitive science are in fact thought to be innate adaptations (though they may be far fewer in number than the hundreds posited by some evolutionary psychologists). Similarly, there may be some innate cognitive capacities that are not full-blown modules, but few philosophers and cognitive scientists nowadays think that there will be isolated innate concepts. If innateness pertains primarily to innate cognitive capacities, then I have argued that they will tend to have a cluster of features in common (e.g. canalized, not learned, informationally encapsulated, and so on), given what we know about human cognition. When Fodor posited modules, he also associated them with other properties that I have not associated with innate cognitive capacities (e.g. fast, automatic, mandatory), though perhaps a weak causal link with some of these properties may also be discovered. Be that as it may, there will be exceptions to most of these properties in any given instance and the clustering of these properties will not be perfect. In explicating the kind *innate cognitive capacity*, I have drawn the boundary around a range of properties that are often, but not always, co-instantiated, and one could draw it more strictly around a smaller subset of properties that are more strongly associated with one another. But then one would miss some causal processes and neglect a number of (exception-ridden) empirical generalizations.

A third objection would question the comprehensiveness of the account that I have given, asking whether there may not be other features associated with innateness that I have failed to consider. In response, it bears repeating that there are various alleged features of innateness that I have already argued do not pertain properly to it, for example, several of those mentioned by Samuels, such as being present at birth, adaptive, and monomorphic. But that obviously does not rule out the existence of other properties that ought to have been included in my account but were not. Mameli and Bateson (2006) list twenty-seven features commonly associated with innateness (in order to argue that none of them ought to be identified with innateness), but most of these features are either ones that I have included or ruled out. The latter category includes various properties that only pertain to the biological rather than the cognitive domain, or those involving genetic determination (e.g. being genetically encoded or genetically influenced), since I have argued that there need be no genetic mechanism that is responsible for the properties associated with innateness. There may yet be other features that I have not considered, which may prove to have causal links to the network I have elaborated. It is difficult to say with finality that no other properties are involved, but if there are other properties then I submit that that would tend to enrich the causal network and strengthen this account of innateness.

3.6 Conclusion

The category of innateness has been criticized as being incoherent and inappropriate for use in a mature cognitive science. In this chapter, I have tried to argue that at least some of the properties associated with the category of innateness are causally linked in a manner that is generally characteristic of real kinds. In Chapter 1, I argued that kinds in the cognitive sciences are validated by the role that they play in causal networks, like kinds in other sciences, both basic and special. When the instantiation of a property or, more commonly, the co-instantiation of a cluster of properties leads causally to the instantiation of a multitude of other properties in recurring causal processes, we identify such a property or set of properties with a natural or real kind. At the current juncture in cognitive science, innate cognitive capacities seem to fit this template well. I also argued in this chapter that there is not likely to be a single type of genetic or neural mechanism or process that underlies all innate cognitive capacities, which suggests that the cognitive kind *innateness* is multiply realizable relative to neural kinds. Moreover, given the nature of the properties associated with the kind *innateness*, some of which are relational or etiological in nature, there is also likely to be a many-to-one relationship between this cognitive kind and its neural correlates. This reinforces one of the central claims in this book, namely that in at least some cases, the relationship between cognitive kinds and neural kinds is likely to be many-to-many.

CHAPTER 4

Domain Specificity

> It is also very worthy of remark, that, though there are many animals which manifest more industry than we in certain of their actions, the same animals are yet observed to show none at all in many others ...
> – René Descartes, *Discourse on Method*

> Evolution behaves like a tinkerer who, during eons upon eons, would slowly modify his work, unceasingly retouching it, cutting here, lengthening there, seizing the opportunities to adapt it progressively to its new use.
> – François Jacob, "Evolution and Tinkering"

4.1 Introduction

There are many contexts in cognitive science in which it is useful to distinguish domain-specific cognitive capacities from domain-general ones, and the distinction seems to do some explanatory work. For instance, according to many evolutionary psychologists, human cognition consists largely of domain-specific cognitive capacities, and this feature of our cognitive makeup provides evidentiary support for the pervasive influence of evolutionary processes on the formation of the human mind (e.g. Carey & Spelke 1994; Cosmides & Tooby 1994). By contrast, according to other researchers, domain-general cognitive abilities are the norm in the human mind and are what distinguish human cognition from that of most other animals (Samuels 1998; Fodor 2000). Yet, the distinction between the two kinds of capacities, domain-specific and domain-general, is not easily drawn. Moreover, some of the examples that are put forward to illustrate the distinction seem to be either spurious or misleading. Like innateness, domain specificity is commonly thought to be a feature of cognitive capacities, processes, systems, and related cognitive kinds. It tends to be applied to capacities that are restricted or constrained in their range of operation in some way. The challenge is to spell out the nature of this restriction and to demonstrate that all such restricted cognitive capacities have something substantive in common that would warrant treating them as members of a kind.

One of the main challenges in explicating the cognitive construct of domain specificity has to do with the difficulty of defining domains in the context of cognition. Researchers in this area regularly distinguish domains like: physical objects, language, number, morality, and theory of mind. But the boundaries of these domains are very permeable and it is not immediately apparent what it would mean for a cognitive capacity to be restricted to one of these domains, or indeed, what it would mean for it to be able to range across (or beyond) one of them. Thus, one of the tasks of this chapter will be to address this difficulty and determine whether it can be satisfactorily resolved. Another challenge is that the term "domain-specific" appears in a number of guises in cognitive science, and some uses seem more tenuous than others, or more aptly captured by other expressions. In what follows, I will focus on what I take to be the most influential usage, though I leave open the possibility that there are other uses that are also worth preserving (preferably using other labels).

In this chapter, I will begin in Section 4.2 by relating domain specificity to some other cognitive constructs, including the constructs of *modularity* and *innateness* (the latter of which was examined in Chapter 3). In Section 4.3, I will propose an account of the phenomenon of domain specificity based on a paradigmatic example as well as on existing theoretical proposals in the cognitive science literature. I will test this account of domain specificity on additional examples in Section 4.4, bringing out some of the distinctive features of this understanding of domain specificity, and attempting to determine whether the category of *domain specificity* corresponds to a cognitive kind. In Section 4.5, I will respond to a theoretical challenge to characterizing domain specificity, which has been termed the "grain problem," before concluding in Section 4.6 that domain specificity can be considered a cognitive kind when suitably described.

4.2 Domain Specificity and Its Confounds

Domain specificity is a feature of cognitive capacities that is often associated with several other such features, notably: modularity, innateness, and brain localization. In this section, I will examine the connections that may or may not exist between domain specificity and these other features, in order to gain a preliminary understanding of the cognitive category of *domain specificity*. By virtue of the way in which *modularity* was initially defined by Fodor (1983), there is a strong link between modularity and domain specificity. Indeed, it follows from Fodor's account that domain specificity is one of the defining features of modularity, and therefore that all modular cognitive capacities are domain-specific.[1] Of course, a case might be made for rejecting this definition on the grounds that it is unwarranted by the empirical facts or otherwise detrimental to research in cognitive science, but I will not try to make that case, nor do I think that the case can easily be made. Fodor's definition appears to have been widely accepted by cognitive scientists, including those who reject modularity, and I will not try to oppose it here. Hence, I take it as uncontroversial that domain specificity is one of the characteristic features of modularity, though the two concepts are not identical and ought not to be conflated.[2]

Things are more complicated when it comes to *innateness*. Although there is also a widespread assumption that there is a link between innateness and domain specificity, there is no prima facie reason for inferring such a link. It is not obvious that all innate cognitive capacities are domain-specific, nor that all domain-specific cognitive capacities are innate. To illustrate, human beings may have an innate cognitive capacity for associative learning that may be entirely domain-general. Conversely, there may be certain cognitive abilities that appear to be domain-specific that are not innate but

---

[1] In addition to being domain-specific, according to Fodor (1983), modular cognitive capacities are supposed to: (2) process items automatically and in a mandatory manner, (3) be inaccessible to consciousness, (4) be fast, (5) be cognitively impenetrable (e.g. resistant to being unlearned), (6) process "shallow" or highly salient features, (7) have fixed neural architecture, (8) have specific breakdown patterns (as in aphasia, agnosia), and (9) have fixed ontogeny (standard pace and sequence of development). Many subsequent discussions take domain specificity to be one of the most central features of modularity. For example, Sperber (1994, 40; emphasis added) writes: "The rough idea of modularity is also clear: A cognitive module is a genetically specified computational device in the mind/brain (henceforth: the mind) that works pretty much on its own on *inputs pertaining to some specific cognitive domain* and provided by other parts of the nervous systems (e.g., sensory receptors or other modules)."

[2] For the sake of completeness, I should mention two caveats. The first is that Fodor restricted his modules mainly to input-output cognitive mechanisms (such as perceptual and sensorimotor capacities), whereas subsequent theorists have posited that modules are more prevalent and may include a range of more central cognitive systems. The second concerns the precise nature of the connection between modularity and domain specificity. Fodor proposed the nine features of modularity in the spirit of necessary and sufficient conditions, but it may be more plausible to regard them as a looser cluster of features that are usually associated with modularity, though perhaps no single one of them is necessary.

4.2 Domain Specificity and Its Confounds

mainly learned, such as chess-playing ability.[3] Having said that, we will have to revisit this link between domain specificity and innateness in the next section, after I have provided what I take to be the most defensible account of domain specificity in cognitive science. It will turn out that on that account, domain specificity only applies to innate cognitive kinds.

As for the link between domain specificity and *brain localization*, this is also widely made, as is the link between brain localization and modularity (which, as seen above, subsumes domain specificity). However, there does not seem to be a cogent reason for making either link. For instance, there are good grounds for thinking that various psychological capacities are modular (and hence domain-specific) even though they are not localized in one region of the brain, indeed even though they are scattered across a range of brain regions (vision and language are obvious examples). Modularity and domain specificity pertain largely to the functioning of a cognitive capacity rather than its neural implementation, so there is limited scope for inferring brain localization from either of these phenomena. Perhaps part of the reason for a conflation of the two notions is that the adjective "domain-specific" is sometimes loosely applied to a cognitive function C when it has been demonstrated that there is a *specific* neural mechanism N or a *specific* brain region R that subserves C (which does not subserve any other cognitive function, C^*). But there are other, more standard ways of referring to such a relationship, namely *selectivity*, or by saying that N is the *neural correlate* (or *neural substrate*) of C. Hence, use of the term "domain specificity" would appear to be misplaced in this context, and it is an unfortunate coincidence that the term "specificity" is sometimes used to denote neural specialization.[4]

In distinguishing domain specificity from other features of human cognition, I have so far relied on an implicit preliminary understanding of the phenomenon. As the term implies, what it is for a cognitive capacity to be domain-specific is for it to pertain to a single domain or to a restricted range of domains, and more importantly, for it not to be generalizable to

---

[3] In Khalidi (2001), I argued that when it comes to domain-specific abilities, it is easier to tell whether and to what extent they are innate or not. That is because we can more easily gauge the amount of explicit learning or relevant experience in the case of domain-specific cognitive capacities than in the case of domain-general ones. It is simply easier to rule out relevant sources of information in the former case than in the latter. But this does not mean that domain-specific abilities are more likely to be innate, just that the evidence is easier to assess.

[4] Here is a recent instance of this conflation from a paper in social neuroscience: "The brain is certainly not equipotential. However, there remain a number of interesting and difficult questions about the degree of such apparent *specialization*, how it might come about and what it accomplishes. These issues have been the focus of numerous theories of *domain specificity*, which range from abstract cognitive hypotheses to neurophysiological and neuroanatomical accounts" (Spunt & Adolphs 2017, 559; emphasis added). The purported link between domain specificity and neural localization will be revisited in Section 4.4.

other domains. Moreover, this last proviso highlights the importance of reserving domain specificity for aspects of cognition that are in principle *generalizable* across domains, although they are not in fact *generalized*. The idea is that a cognitive construct should be generalizable in principle, in the sense that it should be the kind of thing that could be generalized. It would be vacuous to describe as domain-specific some cognitive entity that is not even in principle generalizable. For example, a body of information pertaining to some domain or another (e.g. a list of world capital cities, or the entries in an address book) is not generalizable, since its subject matter is in principle restricted to a certain domain.[5] By contrast, a rule that is deployed by a cognizer in one domain but that *could* be deployed in another domain is in principle generalizable (e.g. modus ponens). Hence, we will mainly be concerned with domain-specific rules, principles, or algorithmic processes. These are the kinds of cognitive entities that can be generalized across domains and that are therefore candidates for being domain-specific, though they may not be the only cognitive capacities that can be so generalized.[6] Derivatively, cognitive capacities or systems can be characterized as being domain-specific when they comprise or include one or more domain-specific rules or processes. In the rest of this chapter, I will be speaking mainly of domain-specific capacities, but also sometimes, of rules, principles, or (algorithmic) processes. The next section will look at a particular example of a domain-specific capacity and use it to advance a proposal as to how to characterize domain specificity.

4.3 A Preliminary Example and a Theoretical Proposal

It will be helpful to begin with a relatively uncontroversial example of domain specificity, though I will argue that it is somewhat misleadingly described and improperly contrasted with a putative case of domain generality. Cosmides and Tooby (1994) explicate the well-known example of the alarm calls of vervet monkeys, who give three different calls in response to three different kinds of predators (leopard, eagle, and snake), leading conspecifics to take three different types of evasive action (respectively,

---

[5] There may seem to be a fairly simple sense in which one body of information is more generalizable than another, for example, a database containing all the residents of New York state, as compared to one containing all the residents of New York state. But this does not mean that the first database is more generalizable, just that it is more general or comprehensive.

[6] Compare Barrett (2018, 4): "One can then define the domain specificity of a process as the degree to which its operations vary across domains. A perfectly domain-specific process is one that operates only in a single domain and in no others. A perfectly domain-general process is one that operates identically across all domains."

climbing a tree, looking up or diving into bushes, and standing on hind legs and looking into the grass) (cf. Cheney & Seyfarth 1990 cited in Chapter 1). In this case, they state: "A single, general-purpose alarm call (and response system) would be less effective because the recipients of the call would not know which of the three different and incompatible evasive actions to take" (Cosmides & Tooby 1994, 89–90). The problem with this observation is not that there could not be a general-purpose alarm system; there clearly could. But a general-purpose alarm system is *not* one that would issue the same call for every predator. That would be a system that fails to discriminate among different stimuli. Rather, an all-purpose alarm system would be one akin to a human linguistic alarm system, which issues a different linguistic warning in the case of different predators. There would clearly be certain advantages to such a system, since it would be capable of handling a much wider range of predators or dangers (e.g. "Lion!," "Hawk!," "Stampede of elephants!") and of being made more precise in various ways (e.g. "Tiger to the right," "Eagle to the northwest," "Human with weapon right behind you"). However, it may also involve certain disadvantages, since given the diversity of inputs and outputs, it may take more processing time to issue the correct alarm, there may be more opportunity for error in both transmission and reception, and the evasive action involved may have to be figured out from scratch by the respondent once the alarm is sounded. Determining which of these two alarm systems, the domain-specific vervet system or the domain-general human system, is more efficient and adaptive is not an easy matter. It will clearly depend on various contingencies such as the types of predators typically encountered, the seriousness and urgency of the threats they pose, and other features of the environment. Cosmides and Tooby may ultimately be right that in certain circumstances a domain-specific system may be superior to a domain-general one. But they do not appear to have drawn the distinction between domain specificity and domain generality in the right way.

This example is instructive since, once it is modified in the way that I have just done, it seems to provide a fairly clear contrast between a domain-specific and a domain-general cognitive capacity. The first feature that can be gleaned from the vervet monkey alarm call system is that some cognitive systems for alarm calls are at least in principle generalizable. That is to say, even though the vervet alarms are only issued for a small set of specific predators, it is not hard to conceive of a different alarm system that would extend to other predators (or indeed, to other types of stimuli). Hence, it seems safe to conclude that for one to speak meaningfully of a domain-specific cognitive capacity, it must have the following feature:

(DS1) A *domain-specific* cognitive capacity is one that is in principle generalizable to new domains.

This condition may appear vacuous, but it is designed to rule out cognitive capacities that consist of a body of information or database rather than rules or algorithmic processes, as mentioned in the previous section. Domain specificity, to be meaningful, must be a feature of a cognitive capacity that is at least potentially generalizable. Though this point may seem obvious, the attribute of domain specificity is often conferred on bodies of knowledge or sets of concepts possessed by subjects that are not obviously generalizable, such as knowledge of animals as opposed to artifacts. The question of domain specificity would seem to be at issue only if those concepts are implicated in one or more inferential rules or principles, as when a set of concepts is embedded in a broader theory (along the lines of the theory theory of concepts encountered in Chapter 2). For example, there is evidence that animals are conceived of as having causal "essences," whereas artifacts are conceived of in terms of their function (Gottfried & Gelman 2005; Greif, Kemler Nelson, Keil, et al. 2006; cf. Boyer & Barrett 2005, 102). If so, then the inferential rules or principles associated with each theory might either be domain-general or -specific, depending on whether they are generalizable or not. Otherwise, it is not clear what sense to attach to the claim of domain specificity (or generality).

The first proposed feature of domain specificity makes reference to "new domains," which is a notion that is in need of further explication and justification. A new domain need not be what we might regard as an entirely disparate area of inquiry. Indeed, the underlying problem is that there are no ready-made boundaries that could serve to delimit domains. In the context of a discussion of whether human creativity is domain-general or domain-specific, Sternberg (2009, 25) poses the problem quite compellingly:

> The greatest challenge in understanding the domain generality versus specificity of creativity is in understanding the concept of a domain itself. Is literature a domain, or German literature, or modern German literature, or modern German literature in its original language, or what? Is cognitive psychology a domain, or psychology, or behavioral science, or social science? Because no consensual definition of a domain currently exists, it is impossible at this time to have a clear sense of exactly what domain-specificity means.

In fact, it is not clear that it will ever be possible to come up with a definition of a domain in the context of cognitive science, since knowledge or information does not come neatly divided into delimited parcels. In

4.3 A Preliminary Example and a Theoretical Proposal 107

the case of the vervets, the original domain is thought to be something like: predators commonly encountered by vervets in the wild. An alarm system could *in principle* be generalizable to include the new domain: all predators, or even, all threats. The vervet alarm system is domain-specific because it appears to fail to generalize to these new stimuli. But these new stimuli do not, strictly speaking, have to be drawn from what we would normally consider to be another domain, such as a new sensory modality or a new area of inquiry. At this point, it might be asked, by virtue of what are they to be considered stimuli pertaining to a genuinely new domain? They must at least be stimuli that the cognizer has not encountered before. But that condition is surely too weak, since the domain-specific vervet alarm system clearly generalizes to new exemplars of leopards, eagles, and snakes, which the individual has not encountered before, indeed ones which perhaps no vervet monkey has encountered before. Rather, in this context, a plausible understanding of new stimuli is that they are ones that the system was not originally designed to cope with. This is admittedly a vague formulation and brings in thorny evolutionary considerations concerning the original design or *proper function* of an evolved trait (Millikan 1989; Neander 1991). Though it is not always easy to determine what the proper function of a cognitive capacity is, some reference to it seems inevitable, since cognitive capacities have evolved to fulfill a certain function. Accordingly, their generalizability consists in being able to extend beyond that original function to encompass cases that they were not designed to cope with, or ones that are not normally encountered in the environment in which they evolved. A similar conclusion has been reached by Boyer and Barrett (2005, 98), who write: "The domain of operation of the system is best circumscribed by evolutionary considerations." Barrett (2018, 6) expands on this conception, as follows:

> The actual domain of a computational process is the set of inputs for which the process can or will produce outputs whereas the proper domain is the set of inputs for which the process has been selected to produce outputs. This is roughly akin to the distinction between "selected effect" and "causal role" functions in biology: the selected effect function of a biological trait is the function it was selected to carry out (in the past) whereas the causal role function of a trait includes all of the functions it can in fact perform, whether or not it has been selected to do so …

On this evolutionary conception of domain specificity, a domain-specific capacity or process is one whose actual domain coincides with its proper domain, whereas a domain-general one is one whose actual domain outstrips its proper domain.[7] Not only does the evolutionary understanding of domain specificity allow us to delimit the boundaries of domains in a principled way, I will go on to argue in the rest of this section and the next that it enables us to make sense of the way in which the category of *domain specificity* is deployed in various areas of cognitive science, and results in a good candidate for a cognitive kind.

Based on the considerations just vetted, I propose that the second crucial feature of a domain-specific cognitive capacity is as follows:

> (DS2) A *domain-specific* cognitive capacity is one that systematically fails to yield a correct output, or fails to yield an output at all, in the case of inputs that the capacity did not evolve to deal with.

The need for this second feature can be further justified by reflecting on appropriate examples from the literature. One such case is provided in a classic paper by Cheney and Seyfarth (1985, 197), who describe the domain specificity of certain cognitive capacities in vervet monkeys, as follows:

> Within the social group, the behavior of monkeys suggests an understanding of causality, transitive inference, and the notion of reciprocity. Despite frequent opportunity and often strong selective pressure, however, comparable behavior does not readily emerge in dealings with other animal species or with inanimate objects.

In this example, both features outlined above are clearly in evidence. First, the principle of causality and the rule of transitivity are clearly applicable outside the realm of social interaction with conspecifics. The transitivity rule can be used to infer hierarchy relations among vervet monkeys (e.g. if A ranks higher than B and B ranks higher than C, then A ranks higher than C) but it can also be used to infer information about size, quantity, and other matters (e.g. if object A is larger than object B and B is larger than C, then A is larger than C). However, despite the clear applicability of this rule to domains that go beyond social interactions with conspecifics, Cheney and Seyfarth claim that vervets do not so apply the rule. Second, it is thought that vervets do not use the rule of transitivity on other species or inanimate objects simply because they evolved the rule to deal with the restricted domain of social interaction with conspecifics, which may have

---

[7] For a precursor to this construal of domain specificity, see also Sperber (1994) on the "actual domain" as opposed to the "proper domain" of a module. Sterelny (2003, 190) also delimits domains in terms of adaptation: "Domains correspond to related sets of adaptive problems environments pose for agents; problems which must be solved if the agent is to survive and reproduce." But he does not appear to explicitly characterize domain specificity and generality in terms of proper domains and actual domains.

been a more pressing adaptive problem. In this case, it may seem obvious that interactions with other animal species and with inanimate objects constitute genuinely new domains. There may not appear to be a need to use the second feature of domain specificity to justify the judgment that it does not generalize to genuinely new domains. But it is not a given that social hierarchy does not constitute a domain that also comprises other species, and that any thinker who could apply such a rule to one's own conspecifics could also apply it to members of other species. Hence, (DS2) can be used to confirm that this is indeed a case of domain specificity, since the actual domain of the cognitive capacity does not exceed its proper domain, which in both cases is social interactions with conspecifics.

4.4 Further Evidence

So far, the examples considered derive primarily from cognitive ethology, specifically studies on other primates. But the concept of domain specificity has also had considerable influence in cognitive neuroscience and developmental psychology. I will now consider whether the notion as I have characterized it can be pressed into service in other areas of cognitive science, particularly when it comes to humans.

There is a well-established body of evidence indicating the existence of category-specific semantic deficits in a range of patients with brain lesions and other neural abnormalities. However, the correct interpretation of this evidence remains a source of contention. Caramazza and colleagues have interpreted this evidence as indicating that semantic information is "domain-specific" (Caramazza & Shelton 1998; Caramazza & Mahon 2003). Other researchers have adopted different models to explain some of the same findings. Tyler and Moss (2001) hold that the selective deficits are an emergent phenomenon. Even though concepts are represented in a unitary distributed system, different types of concepts are structured differently. Since concepts in different domains have different internal structures, impairment of brain function leads to their being differentially affected (Tyler & Moss 2001). On the face of it, much of this evidence, and the surrounding debate, seems to pertain not to the question of domain specificity but rather to that of brain localization. When damage to a certain part of the brain results in selective impairment in naming animals but not plants or body parts, the question is whether this is evidence that representations underlying our semantic information concerning animals is localized in a particular area of the brain, or whether they are not localized but that some of them are more impaired than others by such damage.

Although this is an important question in its own right, it does not bear directly on domain specificity as such, as I have already argued (see Section 4.2). Similarly, neuroimaging data that has been brought to bear on this controversy is more pertinent to the question of localization rather than domain specificity. Caramazza and Mahon (2003, 358) think that "there clearly does seem to be neural differentiation by semantic category" based on neuroimaging data. But Tyler and Moss (2001, 246) find that: "The most striking aspect of the neuroimaging data is the extent to which living and non-living concepts activate common regions with only small and inconsistent differences between domains." The neuroimaging data is obtained mainly by testing healthy subjects on a variety of tasks (e.g. silent naming, word-picture matching) and then using various techniques (fMRI, PET scans) to determine whether different areas of the brain are differentially involved when processing content derived from different domains (e.g. animals, tools, food items). But this does not seem to enable us to draw conclusions regarding whether our capacity to think about such domains involves abilities that are generalizable or not. If our knowledge of animals activates different brain areas than our knowledge of tools, that does not mean that any cognitive abilities that range over such domains are restricted to these domains and cannot be applied to others.

At this point, it may be objected that there is at least an indirect connection between domain specificity and brain localization. If domain specificity is understood in evolutionary terms, as proposed in the previous section, then domain-specific cognitive capacities are adapted for a certain function. Given the dependence of cognition on the brain, it is likely that neural hardware would have been dedicated to carry out this function, which implies a certain degree of localization. Hence, the evidence from lesion patients and from neuroimaging studies bears on the question of domain specificity in the sense that localized capacities are more likely to be domain-specific (and domain-specific capacities are more likely to be localized). If we find that different regions are recruited in tasks involving, say, animals and artifacts, then it is likely that the capacity to reason about animals is indeed domain-specific, as is the capacity to reason about artifacts. But this inference is not warranted, for a couple of reasons. First, even though I have argued that all domain-specific capacities are adaptive, it is not the case that all adaptive capacities are domain-specific. Hence, adaptive cognitive capacities that receive dedicated neural resources are not necessarily domain-specific. Second, evolution exploits existing resources, and there is not likely to be dedicated neural hardware for every evolved cognitive function, as emphasized by recent work on neural reuse (see

4.4 Further Evidence

Chapter 1 and references therein). The localization of certain cognitive capacities does not seem integrally linked to the question of whether they are restricted in their application, which is the notion of domain specificity being explicated in this chapter.

Among developmental psychologists, there are some longstanding debates concerning the domain specificity of our cognitive capacities. For example, as seen in earlier chapters, Carey and Spelke have argued that children have innate systems of knowledge that apply to distinct sets of entities and phenomena. Moreover, they add that the domains of human knowledge, such as knowledge of language, physical objects, and number, center on distinct principles. These "core principles" serve to distinguish one domain from another. But despite the fact that Carey and Spelke hold that our cognitive makeup consists of distinct domains, they also claim that conceptual change in these domains occurs in part by constructing mappings between these domains. For instance, mappings between the domains of physics and number play a role in children's reconceptualization of matter and material objects. Though the mapping is slow and difficult, children eventually succeed in using this mapping from one domain to another to differentiate the concept of weight from the concept of density (Carey & Spelke 1994, 191–192). But if one can transplant certain inferential principles from one domain to another, then those principles are likely not domain-specific. As I have already mentioned, it would be misguided to argue that a cognitive capacity is domain-specific merely on the grounds that it pertains to a distinct body of knowledge. The issue of domain specificity does not arise in such cases. Rather, generalizability of rules or principles is key, and in this instance that condition would seem to be satisfied, thus casting doubt on whether the capacities in question are truly domain-specific (though they may still be innate).[8]

Opponents of the claim of domain specificity also sometimes seem to aim their criticism at a different target. Bates (1994/2001) is at pains to distinguish the claim of domain specificity from claims of innateness and (brain) localization. She stresses that a cognitive capacity can have any two of these features without the third. However, her characterization of

---

[8] Here, it should be noted that domain specificity (like innateness, as explicated in Chapter 3) may be a matter of degree. In fact, it might be a matter of degree on (at least) three dimensions. These dimensions can be captured by the following questions: (1) How difficult is it to learn to apply the capacity beyond the proper domain to the actual domain (e.g. in terms of the amount of time required, or number of exemplars)? (2) How restricted is the scope of the actual domain beyond the proper domain? (3) What is the error rate in the actual domain as compared with the proper domain? These would seem to be the main ways in which we might try to quantify degrees of domain specificity.

domain specificity is vague; with respect to language, Bates (1994/2001, 134) says that the claim of domain specificity is that "localized language abilities are discontinuous from the rest of mind, separate and 'special' ..." Moreover, despite her explicit cautionary notes, in presenting the arguments for and against domain specificity, she sometimes implicates innateness or brain localization instead. For example, she argues against the domain specificity of language on the grounds that the brain systems that support language show an extraordinary degree of neural plasticity (Bates 1994/2001, 139). She also characterizes the controversy over the domain specificity of language as follows: "Have we evolved new neural tissue, a new region or a special form of computation that deals with language, and language alone?" (Bates 1994/2001, 138) But, as already stated, that question does not have a direct bearing on the issue of whether knowledge of language or the capacity to learn language can be generalized to other domains. Whether or not there is a brain region that has evolved to deal with language alone concerns innateness and brain localization rather than domain specificity.

Another case for testing this account of domain specificity can be drawn from research on face recognition, perhaps one of the most widely discussed cognitive capacities in this regard. Researchers tend to be divided as to whether the human capacity to recognize the faces of conspecifics is a domain-specific capacity, or whether it is a capacity that is acquired as a result of more general cognitive processes, of the type used to acquire expertise in other areas of human cognition. Without trying to rehearse the voluminous evidence involved, I will mention just two findings that are pertinent to the issue of domain specificity. Humans do not develop expertise for recognizing the hands or bodies of conspecifics that is at all comparable to their expertise for recognizing their faces, as measured by accuracy and reaction time (McKone, Kanwisher, & Duchaine 2007, 12). Similarly, humans show decrease in accuracy in identifying faces when those faces are inverted but do not show such a decrease in identifying the facades of houses in the inversion condition (Yovel & Kanwisher 2004). The capacity to recognize upright faces rapidly and accurately does not seem to generalize to other visual stimuli. Object recognition is a skill that is in principle generalizable to domains beyond faces (e.g. hands, bodies, houses), but it fails to be so generalizable in humans. This is clearly in keeping with (DS1).[9]

---

[9] But see Boyer and Barrett (2005), who review some of evidence against the domain specificity of facial perception.

4.4 Further Evidence

What of (DS2)? Though it is not always explicitly mentioned by researchers who work in this area, I propose that the evolutionary clause is at least implicitly assumed. Consider the following scenario. Suppose it were found that humans could indeed generalize their face recognition capacities to encompass the faces of dogs. Would this show decisively that the capacity is domain-general, after all? Proponents of domain specificity might not give up on their claim that this capacity is domain-specific, insisting that it is a domain-specific capacity that is specific to the domain of faces in general, or perhaps mammalian faces. Indeed, even if further evidence came to light suggesting that this capacity extends to other objects like the facades of houses, they might continue to posit that it is a domain-specific capacity dedicated to the detection of objects with certain salient parts in particular configurations. What would rule out such a challenge? As I argued earlier, there are no ready-made domains that would enable us to dismiss it in principle. Rather, it seems natural to say that such hypothetical data would not be evidence of domain specificity (across a broader domain) because of *evolutionary considerations*. Since it is likely that such a cognitive ability would have evolved to detect human faces rather than, say, faces of humans and dogs (given the relative recency of the domestication of dogs[10]), let alone the facades of houses, any extension beyond the domain of human faces is indeed a generalization of this ability, and an indication that it is not truly domain-specific. In fact, this is explicitly acknowledged by proponents of domain specificity in this area of research. McKone, Kanwisher, and Duchaine (2007, 12) hold that the domain-specific theory "proposes that a face template has developed through evolutionary processes, reflecting the extreme social importance of faces."

A final case often discussed in this regard is one already mentioned in the previous section, namely the finding that human thinkers, starting from an early age, make different categorization decisions and inferences when it comes to living things and artifacts. As mentioned in Chapter 2, even preschool children make a distinction between animals and artifacts, asking different types of questions when confronted with unfamiliar exemplars from each category (Greif, Kemler Nelson, Keil, et al. 2006). Many studies claim that children are "psychological essentialists" when it comes to living things, prioritizing internal properties and hidden causes, but not when it comes to artifacts, where they tend to rely on functions (e.g. Gelman 2004; Gottfried & Gelman 2005). This is usually interpreted as indicating that

---

[10] There is ongoing controversy about the dating of the domestication of dogs, but upper estimates suggest that it occurred less than 30,000 years ago.

they have domain-specific inferential rules that are applied to the domains of living things and artifacts, respectively (see Sloman, Lombrozo, & Malt 2007). In this case, the relevant rules are potentially extendable beyond each domain, yet are usually not so extended. For instance, though children could make inferences about artifacts based on their internal parts, or about living things based on their function, they generally refrain from doing so. Here, it may be thought that domains are well-defined and there is no need to cite evolutionary considerations in determining that these rules are indeed domain-specific. It seems that we may not need to lean on natural selection to make the claim that human thinkers have domain-specific capacities to reason about living things and artifacts, respectively. However, it is not so obvious when one considers evidence that does not fit neatly with this picture. For example, Keil (1996) suggests that the inferential principles associated with living things may sometimes be applied to complex artifacts, such as televisions, cars, and computers. Does this mean that they are domain-general after all? Not necessarily. Evolutionary considerations suggest that what may happen in these cases is that certain outward features of these artifacts (e.g. movement, interactivity) may "trick" the cognitive system into categorizing them as living things, at least in some contexts, and into applying the essentialist principles to them. Moreover, in many such cases, this leads to incorrect inferences concerning these artifacts, which means that systematic errors are made regarding them. That is why these principles remain domain-specific even though they may occasionally be extended beyond the boundaries of the domain of living things. It is not that the actual function of these essentialist principles exceeds their proper function, but that they rely on certain outward features to perform their proper function, effectively leading them to miscategorize some instances and make incorrect inferences. This analysis is somewhat speculative, but what matters is not that it is right but that questions of domain specificity or generality seem to be resolvable only against a background of evolutionary considerations.

Having tested the proposed characterization of domain specificity on a few examples, we are in a better position to briefly assess domain specificity as a cognitive kind. In keeping with the approach of this book, the status of *domain specificity* as a cognitive kind depends on its causal profile, both synchronic and diachronic. Focusing, first, on the etiology of domain-specific capacities, as characterized here, they are all such that they have been selected for some cognitive task and are restricted in their application beyond that task. That is, their actual function does not exceed their proper function. This means that they are individuated in part on the basis of their phylogenetic history. For this reason, Sperber (1994, 51) points out: "The domain of a module is ... not a property of its internal structure (whether described in neurological or in computational terms)." Second, given that domain-specific capacities do not transcend their proper domains, this means that they lack a certain type of flexibility as compared with domain-general capacities. This is a synchronic causal property of domain-specific capacities and explains why they restrict cognizers in certain ways and do not equip them to deal with novel contexts or environments. Third, domain-specific systems may exhibit a certain kind of efficiency or superior ability when it comes to the particular type of cognitive task that they were selected for. By contrast, domain-general systems, which tend to be more flexible, are also less efficient. As Boyer and Barrett (2005, 100) observe: "In general, the more an inference system exploits external sources of information and stable aspects of the cognitive environments, the more computational power is required to home in on that information and derive inferences from it." This helps explain why domain-specific capacities tend to be fast and automatic (two other features associated with modularity): Since they are dedicated to a specific function and cannot outstrip it, the input–output mappings are likely fixed (which promotes speed) and preprogrammed (which makes them automatic). Hence, this causal profile may also help explain why some of the properties associated with modularity cluster together. This thumbnail sketch of the etiological and causal profile of domain-specific (and domain-general) cognitive capacities brings out some of the features that each class has in common and provides some reasons for thinking that they are genuine cognitive kinds.[11]

4.5 A Theoretical Challenge and Response

The chief theoretical challenge to an evolutionary account of domain specificity has been discussed in depth by Atkinson and Wheeler (2004). On their view, what they call the "grain problem," which is closely related to the problem of delimiting domains encountered above, persists even if we adopt an evolutionary understanding of domain specificity. Moreover, they think that the grain problem for domain specificity is just a special

---

[11] It should go without saying that proper domains need not be "philosophically correct" or "given by reality," as emphasized by Boyer and Barrett (2005). That is, they might not correspond to ontological divisions made by philosophers or scientists. "Faces are distinct objects only to an organism equipped with a special system that pays attention to the top front surface of conspecifics as a source of person-specific information" (Boyer & Barrett 2005, 98).

case of a more general problem with evolutionary explanations that has been noted by Sterelny and Griffiths (1999). Atkinson and Wheeler (2004, 161) formulate the problem as follows:

> Is choosing a mate a single adaptive problem, or is it a set of related problems, such as: choosing someone of the opposite sex, someone who shows good reproductive prospects, and someone who shows signs of being a good parent? Or at a yet finer-grained level of description, is the problem of choosing someone with good reproductive prospects a single problem or a set of related problems, such as choosing someone who is young, healthy, of high status, etc.?

While Sterelny and Griffiths think that this problem can be overcome by adverting to the "device" or mechanism that is of interest in the explanation, Atkinson and Wheeler explain that even devices and mechanisms are subject to a grain problem. For example, in attempting to explain the adaptive function of a leg, do we focus on the adaptive function of the entire leg, or the foot, or a toe, or a bone in the toe? Does each of these have a separate adaptive function, or is it only the foot as a whole that has an adaptive function in its own right? This is why they think the grain problem has two dimensions, involving both the grain of the adaptive problem and the grain of the adaptive solution. Ultimately, Atkinson and Wheeler think that evolutionary explanations of cognition can bring the grain problem under control, but I think that they misrepresent the nature of the solution. To explain why, I will first consider their way out, then say why I think the resources for dealing with the grain problem are already implicit in the account of domain specificity that I have provided.

Atkinson and Wheeler's answer to the grain problem relies on the idea that by moving back and forth between the adaptive problem and the mechanism that is the solution, cognitive scientists and evolutionary psychologists can arrive at a satisfactory answer as to the domain specificity or generality of a cognitive capacity. As they put it: "Ideally, there is a dynamic and mutually constraining relationship between attempts to infer architectural solutions from adaptive problems and attempts to infer adaptive problems from architectural solutions" (Atkinson and Wheeler 2004, 169). They illustrate their solution with reference to research on emotions. This work has shown that "distinct neural structures are disproportionately involved in the perception of fear, on the one hand, and of disgust on the other" (Atkinson and Wheeler 2004, 165). Based on evidence from patients with neural damage as well as healthy subjects, the amygdala has been found to be involved in recognizing fearful facial expressions but not in recognizing expressions of disgust, happiness, and surprise. Meanwhile,

4.5 A Theoretical Challenge and Response

the basal ganglia and insula are recruited in the recognition of facial expressions of disgust, but not of fear. Hence, they claim, there is a double dissociation between mechanisms involved in the recognition of the emotions of fear and disgust.[12] Meanwhile, evolutionary psychologists have provided reasons to think that the perception of fear and disgust emotions evolved under different selection pressures and for different reasons. For instance, there is selective pressure for a specialized system that detects potential danger and threat, and one way of detecting this is by perceiving fear in the faces of conspecifics. Moving back and forth between the implementational level and ecological level, Atkinson and Wheeler think that the grain problem can be finessed. Though they do not say so explicitly, it appears that they are claiming that in this case, at least given the current evidence, there is little difficulty in saying that the capacities to recognize emotions of fear and disgust are domain-specific capacities and that there is no domain-general capacity in humans to recognize emotions in the facial expressions of conspecifics. Both the brain mechanisms and the evolutionary scenario point to the conclusion that these are distinct capacities that have evolved to deal with distinct emotions. First, the fact that the two recognitional capacities are subserved by different neural mechanisms points to a different evolutionary origin for them and hence to the fact that capacities were designed to deal with different emotions. Second, there is a plausible evolutionary story to be told about the need for a separate system to detect threats and to register alarm in the faces of others.

Even though I am generally sympathetic to the conclusion they reach, I will argue that there are two flaws with Atkinson and Wheeler's analysis. First, as I have already argued, the neural implementation of a capacity is indirectly relevant to the question of its domain specificity or generality. At best, they have provided some weak circumstantial evidence to the effect that there is no single (domain-general) capacity to recognize facial expressions of emotion in conspecifics. But it is quite possible that there is a single domain-general system to detect emotion in the faces of conspecifics that is not localized in a single brain region, as opposed to two separate domain-specific systems to detect fear and disgust. Second, the real key to this case lies in the plausibility of the adaptive scenario that Atkinson and Wheeler allude to, which posits a different evolutionary origin for the

---

[12] In fact, contrary to Atkinson and Wheeler's claim, there is substantial evidence against double dissociation in this case, some of which has appeared since their paper was published. For instance, Bonnet, Comte, Tatu, et al. (2015) discuss the role of the amygdala in the perception of positive emotions, and Sander, Grafman, and Zalla (2003) reject the functional specialization of the amygdala. (I am grateful here to Dylan Ludwig for guidance.)

capacities to recognize fear and disgust. If that scenario can be sustained, it would show that there is probably no domain-general capacity to recognize emotion in the facial expressions of other members of our species. But notice that here as in other cases, one could still pose the question (concerning each of these capacities) as to whether there is a single domain-specific capacity that pertains to each of these emotions. For instance, someone could ask whether disgust perception is a domain-specific capacity that pertains simply to disgust, or a capacity that evolved to perceive disgust toward foods, say, and then was generalized to perceive disgust toward a range of other objects (e.g. human actions). The answer to this question would rest on the plausibility of the evolutionary evidence that one could produce in favor of either of the two scenarios. Indeed, when it comes not to the *recognition* of disgust in others but to the *emotion* of disgust itself, some cognitive scientists have posited that the food evaluation and rejection system was later applied to the evaluation and rejection of social action, so that feelings of disgust toward human action were generalized from feelings of disgust toward food (see e.g. Haidt & Joseph 2007). If this evolutionary scenario is itself plausible, and if there is some evidence to suggest that the recognitional capacity coevolved with the emotion itself, then a similar account might be given of our ability to recognize disgust in others (namely, that it initially evolved to recognize disgust toward food and was later generalized). In short, I contend that Atkinson and Wheeler have misrepresented the means of resolving the grain problem in this and in other cases. In each case, what enables us to determine whether a cognitive capacity can be considered domain-specific is an account of its proper function, in other words, the adaptive purpose that it evolved to serve. In accordance with requirement (DS2), if the cognitive capacity enables the organism to carry out tasks that go beyond its proper function, as determined by the most likely evolutionary scenario, then it can be considered domain-general; otherwise, it is domain-specific.[13]

In the previous example, it may seem odd to speak of a domain-general capacity for recognizing expressions of disgust in the faces of others (on the assumption that it evolved to detect expressions of disgust toward food). It may be objected that such a capacity would surely still be domain-specific,

---

[13] Another shortcoming of Atkinson and Wheeler's account is that it considers domain specificity in principle to be an attribute of either information or rules (2004, 151). But I would argue that a body of information that is fully specified and does not contain open variables always pertains to a certain domain, and hence the possibility of generalizability to new domains does not even arise in this case, as required by (DS1). That is why the only candidates for domain specificity are rules, principles, and algorithmic processes.

4.5 A Theoretical Challenge and Response

being restricted to disgust, as opposed to other emotional expressions such as fear. But despite the fact that it may seem counter-intuitive to say that a disgust-detection capacity is domain-general, on the evolutionary scenario that I sketched out, it would indeed be one that has actually been generalized by human cognizers beyond its proper function. Though disgust may seem a restricted domain by comparison with emotion, domain specificity has to do with the cognitive ability to go beyond a certain given domain, not the inherent specificity of that domain itself (indeed, one cannot even make sense of such a notion, in the absence of antecedently delimited domains). Our pre-theoretical intuitions about this and other cases may need to be revised in light of what we discover about the evolutionary history of the cognitive capacities that we examine.

Given its reliance on evolutionary considerations in delimiting domains, the account of domain specificity that I am proposing clearly only applies to evolved cognitive capacities. Since evolved phenotypic features of the organism are at least in part innate, it has turned out on this account that only innate capacities can be domain-specific. Hence, this would provide an integral link between innateness and domain specificity, contrary to what was claimed in Section 4.2, above. If the argument to this point has been correct, it is a link that results from the fact that a definite sense cannot be attached to the notion of a domain-specific capacity unless it is an adaptation of some kind, and hence innate. This may seem like a shortcoming of the account, since at least prima facie it seemed conceivable for a cognitive capacity to be domain-specific yet not innate. But if that hunch is correct, it would require us to find some other way of explicating domain specificity that does not cite evolutionary considerations.[14]

This account of domain specificity in terms of proper function may raise worries about evolutionary explanations in general, especially when it comes to psychological traits. It is difficult enough to reconstruct evolutionary history when it comes to morphological traits that leave physical traces, so it may seem like a thankless task to try to do so in the realm of cognition. Determining whether a cognitive capacity is domain-specific might be thought to be empirically intractable. But there has been a large body of scientific work on adaptive accounts of psychological

---

[14] What about cognitive capacities that acquire functions via their learning histories or ontogenetic development? Can we say whether they are domain-specific or -general, even though they do not have evolved proper functions? Though I will not try to justify this here, it may be possible to generalize the notion of function that I am exploiting to characterize domain specificity to cover such cases, along the lines proposed by Garson (2019) in his "selected effects" account of functions. (I am grateful to an anonymous referee for pressing me on this point.)

capacities and there are diverse pieces of evidence that can be brought to bear in trying to determine the proper function of a cognitive capacity. Ereshefsky (2007; 2012) has made a strong case for considering "psychological categories as homologies" and has outlined various methodologies and strategies that evolutionary biologists, ethologists, and others have for identifying "behavioral homology." This requires reconstructing phylogeny, as well as ontogeny, in an effort to determine the precise lines of descent of certain psychological traits. In fact, Ereshefsky contrasts the homological approach to investigating psychological categories with a functionalist one, on the grounds that functionalist accounts in psychology tend to be ahistorical and privilege synchronic causal roles. But in this chapter and elsewhere in this book, I have interpreted functions at least partly in etiological terms, and that accords with the approach of at least some of the cognitive scientists working on domain specificity (e.g. Sperber 1994; Sterelny 2003; Boyer & Barrett 2005). It is also in keeping with Ereshefsky's suggestion of combining functionalism with a homological approach to psychological traits, by building on the work of biologists who have argued that adaptational approaches to behavior and psychology have neglected the phylogeny and ontogeny of psychological traits (Ereshefsky 2007, 660). This approach and the various methodological strategies used by scientists in ethology, comparative cognition, and other areas, vindicate the feasibility of tracking the etiology of cognitive capacities in an attempt to ascertain their proper functions. As Ereshefsky (2007, 671) puts it: "The lesson for studying psychological categories as homologies is that regardless of our access to the historical record, an array of phylogenetic methods exists for testing adaptational hypotheses." Exploiting these methodological strategies should help us avoid positing just-so stories or falling into the crude adaptationism that characterizes some work in evolutionary psychology.

4.6 Conclusion

In this chapter, I have tried to provide an analysis of domain specificity in cognition that enables us to make a theoretically useful distinction between domain-specific and domain-general cognitive systems. Drawing on examples from the literature, both genuine and spurious, I have tried to show that there are two features that make a cognitive capacity domain-specific. First, the cognitive capacity must be one that is in principle generalizable. Hence, it cannot be something like a body of information concerning a particular area, but something more like a rule or algorithmic

4.6 Conclusion

process that has wider applicability.[15] Second, it must be a capacity that the subject is unable to apply to genuinely new cases, where new cases are ones of a type that this system was not originally evolved to deal with, or ones that are not within what has been termed the *proper function* of this cognitive system. This second condition is important in that it provides us with a principled way of delimiting domains, since these are not antecedently given. Although it is often difficult to determine what the proper function of a cognitive capacity is or what it has evolved to deal with, since that requires us to understand the adaptive pressures that led to its being selected in the ancestral environment, this seems no more difficult in principle than the determination of evolutionary functions concerning other phenotypic features. This is what enables us to rule whether a cognitive capacity is one that the agent can extend to genuinely novel domains or not. The distinction between domain specificity and domain generality matters because a central debate in contemporary cognitive science concerns the extent to which our cognitive capacities are domain-specific tools or whether they are domain-general problem-solving capacities. A resolution of this disagreement depends on a clear means of demarcating domain-specific from domain-general systems or capacities. Moreover, it has often been claimed that one of the main points of difference between human cognition and that of other animals is its domain-general nature. Again, this debate cannot be properly adjudicated unless we have a principled way of making the distinction.

As a cognitive kind, domain specificity is related to innateness in at least two ways. The first is that it is a second-order kind, applying primarily to cognitive capacities or other first-order cognitive kinds. It may even be construed as a *property* of cognitive capacities, rules, or principles, since as I have pointed out at various points in this book, the distinction between kinds and properties may not run very deep, at least in cognitive science. The second way in which it is related to innateness is more direct. As argued in this chapter, according to this characterization of domain specificity, the only candidates for domain-specific cognitive kinds are evolved kinds. They are the ones that have proper functions that have been selected by the process of natural selection. These cognitive kinds are therefore innate to the mind, along with other evolved mental kinds. This means

---

[15] The question may be raised whether this means that domain-specific capacities are pitched at the algorithmic, not computational level. Though these capacities will include algorithmic processes, a judgment of domain specificity does not need to ascertain the precise algorithms involved or identify their main steps. The point is to determine the net effect of the capacity and its function, which is very much something to be determined at the computational level.

that domain-specific kinds are individuated, at least in part, with reference to phylogeny. Like some of the other cognitive kinds discussed (and to be discussed) in this book, it is identified on the basis of its causal history. As such, it does not supervene entirely on the intrinsic properties of the individual thinker but rather on the thinker's causal history and the causal history of the species. Part of what it is for a cognitive capacity to be domain-specific is for it to have a certain phylogeny, so if (to conjure a far-fetched scenario) two cognitive capacities are intrinsically identical they may not both be domain-specific if they do not share the same phylogeny. That is not to say that it will be easy to tell whether a cognitive capacity has the right etiology for it to qualify as a cognitive kind, but a definitive determination entails reconstructing phylogenetic history.

CHAPTER 5

Episodic Memory

If any one faculty of our nature may be called *more* wonderful than the rest, I do think it is memory. There seems something more speakingly incomprehensible in the powers, the failures, the inequalities of memory, than in any other of our intelligences. The memory is sometimes so retentive, so serviceable, so obedient; at others, so bewildered and so weak; and at others again, so tyrannic, so beyond control! We are, to be sure, a miracle every way; but our powers of recollecting and of forgetting do seem peculiarly past finding out.
— Jane Austen, *Mansfield Park*

It's a poor sort of memory that only works backwards.
— Lewis Carroll, *Alice in Wonderland*

5.1 Introduction

Memory is a cognitive category that has a very illustrious history and an entrenched position in our cultural and linguistic practices. *Episodic memory* is a more recent classification, though it is arguably a category that many cultures possessed implicitly before it was explicitly identified by psychologists, even though they may not have distinguished it consistently from what are now widely considered other subtypes of memory.[1] In recent taxonomic practice in the cognitive sciences, a distinction is usually made between *episodic memory* and *semantic memory*, where the former concerns the capacity to retain information from experiences pertaining to events that occurred in one's own personal experience, while the latter involves a capacity to do the same with information of a more generic type that is retained independently of the circumstances in which it was acquired (cf. Foster 2009, 39–40). First explicitly identified and labeled by Tulving

---

[1] However, for an argument that the concept MEMORY is not a cultural *universal*, see Wierzbicka (2007). Some authors claim that what is ordinarily called memory is (roughly speaking) episodic memory – and this is what past philosophers and psychologists also traditionally thought. Klein (2015) thinks the term "memory" should be reserved for episodic memory.

(1972), episodic memory is often what we mean by the word "memory" in English: our capacity to retain and later recall autobiographical experiences, and the mental states produced by this capacity. The rest of this chapter will look at the arguments for and against considering *episodic memory* to be a valid category that corresponds to a real kind in the cognitive sciences. But before adjudicating that debate, it is worth considering the question as to whether the superordinate category *memory* corresponds to a real kind. Various theorists have argued that *memory* does not correspond to a real cognitive kind, so it is worth briefly exploring some of the arguments that have been put forward for this conclusion.

The standard scientific taxonomy of memory now includes a number of sub categories. I have already mentioned the categories of *episodic* and *semantic memory*. These two categories are subsumed under the superordinate category, *declarative memory*. *Declarative memory* is often distinguished in turn from *non-declarative memory*, which itself is subdivided into two or more subtypes, including *procedural memory*, *classical conditioning*, and others. This yields the taxonomic picture illustrated in Figure 5.1 (a), which shows that the over-arching category *memory* is now thought to be subdivided into at least two main types (*declarative* and *non-declarative*), and the latter is in turn subdivided into two main types (*semantic* and *episodic*). It may already seem unlikely that such an apparently heterogeneous category (*memory*) could correspond to a real kind. Given the diversity of the types of abilities and information associated with each cognitive capacities, there may seem to be insufficient commonality among them. Moreover, on some taxonomic divisions of memory, a further distinction is often made between *long-term* and *short-term memory*. Sometimes this distinction is made before the distinction between declarative and non-declarative memory, on the grounds that both of these types are sub categories of long-term memory, as shown in Figure 5.1. (b). But at other times, the distinction between long-term and short-term memory is considered orthogonal to the declarative/non-declarative distinction; in other words, there can be both long- and short-term declarative memory, as well long- and short-term non-declarative memory.

A number of authors have argued that there is no superordinate category of *memory* that plays a useful role in cognitive science. Michaelian (2011b) points out that the different subtypes of memory are now widely considered to correspond to different cognitive *systems*. He proposes to individuate memory systems with reference to Marr's three levels, the computational, algorithmic, and implementational levels, which are foundational for contemporary cognitive science (and which have already been

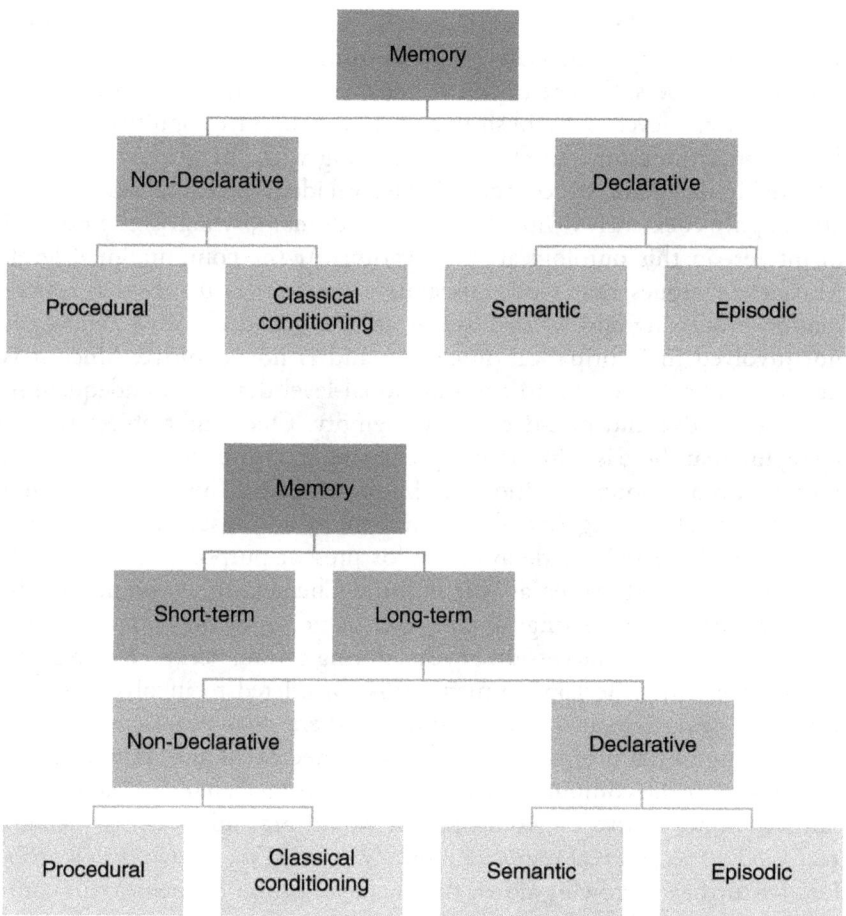

Figure 5.1. Two possible taxonomies of memory: The second incorporates the distinction between long-term and short-term memory and posits that both declarative and non-declarative memory are species of long-term memory.

encountered in Chapter 1, and deployed in individuating the kind *concept* in Chapter 2). That is, a cognitive system is distinguished from another cognitive system based on the properties that it has at *each* of these three levels. Accordingly, he defends the conditional claim that if the "multiple memory systems hypothesis" is correct (i.e. that memory is subdivided into distinct systems roughly along the lines indicated above), then memory is not a natural kind. His basic argument relies on saying that there are important differences between declarative and non-declarative memory at all three Marrian levels, which indicates that they are not subtypes of a single overarching kind, *memory*. By contrast with Michaelian, I think that it would be sufficient to show that memory is heterogeneous at the computational level to establish that it is not a unitary cognitive kind, so I will concentrate on that claim.[2] In keeping with the general approach adopted in this book, the computational level identifies causal systems that are roughly coextensive with the cognitive domain, which is the domain of interest in this ontological investigation. At the computational level, Michaelian argues that while declarative memory is involved in information-processing and is cognitive in nature, non-declarative memory is not involved in information-processing and is noncognitive. Since it is not so involved, there is no computational-level description adequate to both declarative and non-declarative memory. One might object to this by saying that there is information-processing in non-declarative memory, though it is of a nonpropositional or implicit variety. This objection could be elaborated by saying that all cases of memory are "instances when information of the past is made available for present purposes" (Werning & Cheng 2017, 7).[3] However, as Werning and Cheng (2017, 7) point out: "In this minimal sense the rings of trees are memories of the climatic conditions in the seasonal succession of years during certain periods of the past." Thus, even if non-declarative memory is considered to involve information processing (albeit nonpropositional), there does not seem to be an overarching causal profile that can be attributed to all and only instances of memory at the computational level. This minimal characterization pertains to noncognitive phenomena, and would preclude memory being a real kind, at least a real *cognitive* kind.[4] We could try to avoid this problem by further narrowing down the characterization of memory in computational terms by referring to the processes of encoding, storage, and retrieval, which are often thought to be distinguishing characteristics of memory. But Klein (2015, 1) points out that if we take memory to include "any state or process that results from the sequential stages of encoding,

---

[2] Michaelian (2011, 178) acknowledges that it would be possible to take this route, which is implied by his argument: "if my argument below that nondeclarative memory is noncognitive succeeds, then my conclusion that memory is not a natural kind follows even given the less restrictive computational-similarity-only approach."

[3] Some authors give an even more expansive definition of memory: "For our purposes, memory can be defined as experience-dependent modification of internal structure, in a stimulus-specific manner that alters the way the system will respond to stimuli in the future as a function of its past" (Baluška & Levin 2016, 903). (I am grateful to David Colaço for the pointer.)

[4] See also Rupert (2013) for an argument that memory is not a natural kind in the cognitive sciences, which employs a notion of cognitive kind that is very congenial to the one being used here.

storage, and retrieval," that would include virtually all aspects of cognition. In other words, this characterization does not pick out a distinctive category at the computational level. This is not entirely convincing, since even though a preponderance of our cognitive *states* may be subject to encoding, storage, and retrieval, that does not imply that all our cognitive *capacities* perform this role. The reason that this may not be obvious is that our memory capacities interact with a broad range of other psychological capacities, such as perception, inference, and emotion, so their output states are generally the joint product of memory and other systems. That may be why this characterization of *memory* might seem to encompass all of cognition. Therefore, contrary to Klein's contention, it is at least worth considering whether any cognitive system that implements a process of encoding, storage, and retrieval may be considered a memory system, and whether the class of all such systems corresponds to a cognitive kind.

I have presented some recent arguments against the claim that *memory* in general can be considered a cognitive kind comprising a number of subkinds (episodic memory, semantic memory, procedural memory, and so on). These arguments contend that there are no common computational or cognitive properties shared by all of the various alleged types of memory, at least not ones that are not also shared by phenomena that are not instances of memory. But these arguments do not appear to be decisive and there may be room for understanding memory as a general capacity for encoding, storing, and retrieving informational content, or perhaps more convincingly, as a capacity for *specific types* of encoding, storage, and retrieval. Thus, it may be that there is a cognitive kind that includes all the capacities included in Figure 5.1, or it may be that there is a less inclusive kind *declarative memory* that includes both *episodic memory* and *semantic memory*, or it may be that *episodic memory* is not a subkind of some more general type of memory. In what follows, in investigating whether *episodic memory* is a real kind, I will pursue the issue independently of the question of its superordinate and subordinate categories, if any. (For these purposes, we can think of "episodic memory" as a simple rather than a composite term indicating both a species and a genus.)

5.2 What Is Episodic Memory?

In the previous section, I mentioned episodic memory and other types of memory as cognitive capacities or systems, and that is how they are often theorized about in cognitive science. But there is another prominent use of the term "memory" in common parlance and in scientific discourse (at least in English), which corresponds roughly to the *output* of such a system, a type of mental state with informational or representational content. In what follows, when it comes to episodic memory, I will be discussing both of these cognitive categories, the capacity and the state, since I think that there is a close connection between the capacity for memory and its outputs, as I will argue in due course.[5] This also accords with common English usage, since we talk about episodic memory both as a capacity (e.g. "My memory for recent events is not what it used to be") and a state (e.g. "I have a fond memory of eating pizza for breakfast last Sunday"). Moreover, when it comes to the cognitive state of episodic memory, there is a possible confusion between two different cognitive entities. One is the mental state that a cognitive agent is in at the time of recollection or retrieval, the memory produced on a particular occasion, and the other is the item that is presumably held in storage. On a naïve view, these two entities are virtually identical: The item that is retrieved by an agent in response to some external or internal stimulus is commonly thought to be the very same item that has been held in storage since the actual experience. But we will see later that this naïve view, which has been called the "archival view" of episodic memory (Robins 2016a), cannot stand up to scrutiny. Accordingly, I will argue that the primary objects of investigation in memory research are the recollected memory *state*, what is retrieved by a particular agent on a particular occasion, and more centrally, the *capacity* that produces such states.

As indicated in the previous section, the first explicit identification of episodic memory as a distinct mental capacity is usually traced to Tulving (1972), though he himself states that the distinction was already implicit in both psychological research and in the philosophical literature. In that work, Tulving characterized episodic memory as a system that stores information with spatiotemporal and autobiographical content (e.g. "I remember that I met a retired sea captain on my summer vacation last year"). By contrast, Tulving described semantic memory as a "mental thesaurus" that stores information necessary for the use of language (e.g. "I remember that summers in Katmandu are usually quite hot") – though as this brief characterization suggests, it might be better characterized as an encyclopedia.[6]

---

[5] Morris (2007, 30) distinguishes five different senses of memory, roughly as follows: (1) the capacity to encode, store, consolidate and retrieve information; (2) a hypothetical store in which information is held; (3) the information in such a store; (4) the process of retrieving information from a store; (5) an individual's phenomenal awareness of remembering something. Compare Tulving (2000, 36), who mentions all of these senses (roughly) but adds one that does not seem particularly prominent, namely a property of the information in the store.

[6] Compare Tulving, Schacter, McLachlan, et al. (1988, 15): "Episodic memory is concerned with the remembering of personal happenings and experiences that are dated and located in subjective time

5.2 What Is Episodic Memory?

As it stands, this way of making the distinction cannot be quite right, since I remember the date and place of my birth whenever I have to fill out certain official forms, which is information with autobiographical and spatiotemporal content, but that is more plausibly considered a product of semantic memory (similar to my memory of, say, my mother's date and place of birth). Rather, the intended distinction seems to be between memories of an event that are based on a direct experience of the event itself and those that are not. Moreover, according to some researchers this experience is in some sense "relived" when the memory is retrieved. In later work, Tulving went on to characterize episodic memory in terms of "autonoetic" ("self-knowing") consciousness. On this revised view, episodic memory involves a distinctive kind of mental experience, or a "special kind of consciousness that allows us to be aware of subjective time in which events happened" (Tulving 2002, 2). Many other researchers have agreed with Tulving that the phenomenal quality of reliving a past experience is what sets apart episodic memory from semantic memory.

There are several things to notice about this more recent characterization of episodic memory and its differentiation from semantic memory. First, it applies primarily to memory as a state rather than a capacity, since autonoetic consciousness evidently pertains to an occurrent mental state that is entertained by a cognitive agent, rather than a mental capacity of some kind. But it seems possible to use this characterization to identify the associated capacity (at least in part) as one that is disposed to produce states of that kind. Of course, even if *states* of episodic memory belong to a real kind, that does not guarantee that there will be a real kind of *capacity* dedicated to producing such states, but it is at least a live possibility. Second, by grounding episodic memory in autonoetic consciousness, researchers do not usually consider that episodic memory is necessarily ineffable or essentially subjective. Although states of episodic memory are characterized in phenomenal terms, episodic memory researchers like Tulving usually consider that this autonoetic aspect is a property that is amenable to scientific investigation from a third-person perspective, even though any investigation will rely heavily on self-reports by experimental participants.[7] They have developed various experimental techniques to

---

and space, whereas semantic memory underlies acquisition and retention of publicly verifiable facts of the world."

[7] Some researchers think that psychology has its work cut out for it in investigating the phenomenology of episodic memory: "Until we can more consistently and carefully direct our empiricism toward capturing phenomenology, psychological science runs the risk of defining out of existence the very thing(s) that make a memory a memory (and, more generally, a human a human)" (Klein 2015, 25).

try to assess this. As Tulving (1985, 6; original emphasis) states: "one way of measuring autonoetic awareness could take the form of asking people, when they recall or recognize a previously encountered item, whether they *remember* the event or whether they *know* in some other way that it occurred."[8] Other cognitive scientists have adopted this technique, notably in experimental paradigms in which participants are shown lists of words and then later asked whether certain words appeared on the original lists.[9] Admittedly, asking people whether they actually "remember" an item or whether they "know" it in some other way is not a foolproof technique for determining whether they have the requisite autonoetic consciousness. But it would seem to provide a fallible way of accessing the experience of the cognitive agents and the nature of the mental state in question. Third, despite the fact that Tulving and many other memory researchers distinguish episodic from semantic memory, many of them also see close links between the two capacities (and their respective states). In particular, some semantic memories may be strongly associated with episodic memories, specifically concerning the occasion on which one acquired the information. For example, I remember that Pretoria is the capital of South Africa (semantic memory), but I also remember when I first learned that piece of information, namely when I gave the wrong answer in a school general knowledge contest (episodic memory). Even when episodic and semantic memories are not associated in this way, it is plausible that some semantic memories originate in episodic memories, many of which are later forgotten (though this is not to say that semantic memory is just the accumulation of episodes; see Baddeley 2001). For instance, I no longer remember when I first learned that Paris is the capital of France, but I do remember that fact. Similarly, some episodic memories can be strongly informed or supplemented by semantic memories. I might not remember all the details of a particular past experience but can "fill them in" using background information (semantic memories) in trying to recollect them to myself or recount them to others (as will be further elaborated in Section 5.4).

---

[8] The terminology here is unfortunate, since in philosophical contexts "knowledge" is generally defined as justified true belief. In this case, Tulving and others do not intend to suggest that their subjects' belief is true and justified, but that they have surmised it without having the requisite phenomenal experience. To complicate matters, some researchers seem to reverse the terminology: "Recognition might solely reflect a feeling of familiarity ('knowing') or recognition might be verified by recollecting something about the episode when that name or person was last encountered ('remembering')" (Aggleton & Brown 2006).

[9] For example, Roediger and McDermott (1995, 807) explain the method as follows: "Essentially, subjects were told that a remember judgment should be made for items for which they had a vivid memory of the actual presentation; know judgments were reserved for items that they were sure had been presented but for which they lacked the feeling of remembering the actual occurrence of the words."

5.2 What Is Episodic Memory?

Notwithstanding these important connections between episodic memory and semantic memory, there is also an impressive body of evidence that they are doubly dissociated. That is, there are instances in which human thinkers exhibit severe deficits in one without appreciable shortcomings in the other. This double dissociation is supported by the existence of patients with acutely impaired episodic memory and largely spared semantic memory (Tulving 2002, 13–16), as well as by semantic dementia patients who have severe deficits in semantic memory with relative preservation of episodic memory (Irish, Addis, Hodges, & Piguet, 2012). Hence, Tulving's original distinction has been corroborated by subsequent empirical work revealing a double dissociation between the two posited memory systems.

Thus far, we do not seem to have resounding evidence that we are dealing with a real cognitive kind. Despite some apparent confluence with first-person experience and vernacular categorization, *episodic memory*, when characterized in terms of its "autonoetic" phenomenal features, does not seem to have the causal profile found among other cognitive kinds discussed so far in this book. However, even though Tulving's characterization of episodic memory in terms of its experiential properties has gained wide currency, there are other ways of individuating it in the psychological and philosophical literature. One way that is particularly prominent in philosophical discussions (and at least implicitly in psychological research) has to do with the claim that it is a type of mental state that carries information about one's personal past. This way of identifying episodic memory relies not on its phenomenal properties but on its etiology and is often associated with the "causal theory of memory" proposed by Martin and Deutscher (1966). Their theory is based largely on conceptual analysis of the terms "memory" and "remember," informed by various hypothetical scenarios.[10] By examining a number of cases and variations, they emerge with a definition of the vernacular term "remember," whose main tenets are roughly as follows:

> If X remembers something public (e.g. a car accident) or private (e.g. an itch), then the following criteria must be fulfilled:
>
> 1) X represents the past thing (within certain limits of accuracy);
> 2) If it was public, then X observed what he represents; if it was private then it was his;
> 3) X's past experience of the thing was operative in producing a state or successive states in him finally operative in producing his representation.

---

[10] In fact, they propose a theory of memory that is meant to cover not just episodic memory, but semantic memory and procedural memory as well; in their words: "direct memory of events, remembering information, and remembering how to do things" (Martin & Deutscher 1966, 161).

They also add two further clauses to (3). The first relates to cases in which the rememberer is prompted to remember by some clue or hint. In such cases, Martin and Deutscher hold that X's past experience of the thing represented must be operative in producing the state (or the successive set of states) in X that is finally operative in producing the representation. The second concerns the relation of the representation to the experience: The state or set of states produced by the past experience must constitute a structural analogue of the thing remembered, to the extent to which X can accurately represent the thing. The upshot of this view is that episodic memories are states that are causally connected in an appropriate (i.e. non-deviant) way to a representation of an experience in a subject's personal past. This connection is secured by means of a "trace," which Martin and Deutscher (1966, 189) describe as "an indispensable part of our idea of memory," describing it as a "structural analogue of what was experienced." As Michaelian (2011, 332) explains in his exposition of their theory, "the memory trace has to exist all along; and it has to be doing causal work at the time of the remembering," even if the memory is cued in some way. Thus, there is an uninterrupted causal chain between initial experience and subsequent representation that is mediated by means of a trace.

There is something intuitive about this notion of episodic memory and it seems to be in accord with at least many uses of the terms "remember" and "memory" in English. As Martin and Deutscher ingeniously show by using various hypothetical cases, we are loath to consider a current state to be a memory unless it has something like the causal history that they specify. This is also at least the implicit position of many researchers in psychology, who regard a causal link to the past as a crucial feature of episodic memory states, alongside their distinct autonoetic phenomenal properties. Klein (2015, 22) puts it as follows:

> To label a mental occurrence as an act of memory is not simply to be able to show that the experience can be causally traced to events in one's past. It requires that the experiencer have a direct, non-inferential feeling that what is now in awareness is a reexperiencing of events that took place in his or her past.

Even though Klein agrees with many other psychologists that autonoetic phenomenology is crucial to episodic memory, he also takes it as uncontroversial that episodic memories must be traceable back to past experiences. As we shall see in Section 5.3, there are some complications associated with this picture of episodic memory, especially in light of some empirical evidence that seems to undermine the existence of a distinct capacity of episodic memory. But before looking at these difficulties, it is worth asking why a science of the mind should be interested in identifying states of this type and categorizing them separately from other mental states, including states of semantic memory, as well as states of imagination, perception, learning, inference, and others.

We often have occasion in science to type entities based on their etiology or causal history. This holds of real kinds in biology (e.g. *species*), geology (e.g. *sedimentary* and *igneous rocks*), astronomy (e.g. *meteorite*), and other sciences (cf. Khalidi 2021). Moreover, we saw in Chapter 2 that cognitive scientists have good reason to individuate *concepts* in part according to their causes. In many cases, categorizing according to causal history in science is in the service of explanation. This would seem to apply to the case of episodic memory in particular. Suppose I see my next-door neighbor arrive at the door of her house, rummage around in her pockets, then lift up the flowerpot near the door and extract a key from under it, before letting herself in. In the typical case, the best explanation of why she looked for her spare key under the flowerpot is that she remembers having left it there. In other words, she has a memory that represents the key being in that location, and that mental state is *caused* ultimately by a past event in which she placed the key under the flowerpot. But one may object here that this explanation does not need to appeal to a state of memory that can be traced back to an episode in her past. What explains her behavior of looking under the flowerpot is simply a mental state that represents the key being there, whether or not that mental state has the right causal history. After all, she may well look for the key there even if she just imagined having put it there, or dreamt it the previous night, or had a vague hunch. But if we want to explain why she *succeeds* in finding it when she relies on her memory (in the event that she does succeed), then the explanation would need to rely on a causal connection to a past episode. Notice that she need not succeed each time she relies on her memory, she need not even succeed more often than not, but when she does succeed, the causal connection provides the best explanation of why she does. Otherwise, the success of the behavior would be a mysterious coincidence.[11]

---

[11] Of course, in some cases, when accurate information about the past can be gleaned from generic background knowledge, a causal link to a specific experienced event may not need to be invoked. For example, if the neighbor looked for the key under the welcome mat in a context in which there was a widespread convention that keys are hidden there, no such link would need to be posited. Compare what Martin and Deutscher (1966, 176) say about explaining the case of a man who paints a scene from his childhood accurately even though he does not believe it happened: "We talk of the 'only reasonable explanation' in the case described, of course. We do not suggest that in no case would there be an alternative explanation."

This particular explanation is couched in a folk psychological context, but similar explanations occur in the context of scientific psychology. Consider, for example, the well-known Deese–Roediger–McDermott (DRM) paradigm, in which experimenters show participants lists of closely related words (e.g. *table, sit, legs, seat*), then determine how many words are correctly remembered, as well as which words are incorrectly recalled and how frequently. In this paradigm, it is found that participants often incorrectly recall words that are not on the list when those words are closely associated with words on the list (e.g. *chair*). In a seminal experiment, the mean probability of recall of studied words was 0.65, whereas the mean recall of closely associated words ("lures") was 0.4 (Roediger & McDermott 1995). While high, this latter probability was significantly lower than the probability of recall of studied words. This experiment shows that words that were not encountered, yet closely related to the encountered words, were incorrectly recalled at a relatively high rate. Here, what is in need of explanation is the production of lures, not the production of the studied words. It does not even occur to researchers to explain the latter fact, since the explanation is obvious: These words were produced precisely because they were previously encountered. This is not to say that all previously encountered words are correctly recalled in such experimental setups, but they are recalled often, and this is just what we would expect. In fact, as emphasized by Robins (2016a) the explanation of why the lures were produced depends on the explanation of why the studied words were produced: They are semantically related to them.[12] There is no need to explain why most experimental participants succeed in recalling many of the studied words. That is the background against which much of the empirical work on memory proceeds: When items are successfully produced it is generally because they have been previously encountered. It is possible that some of these are simply lucky guesses, but in experimental setups of this type it is vanishingly unlikely that they all are.

---

[12] Robins (2016a, 433) writes: "Misremembering is a memory error that relies on successful retention of the targeted event." She argues that explaining misremembering "involves a critical, but often underemphasized, condition: the rememberer must retain information about the particular past event" (Robins 2016a, 443). There are a number of different ways of elaborating this explanation, for example, some researchers think that reading the words on the list may activate a network of associated words at the encoding stage, while others hypothesize that this network activation occurs at the stage of retrieval (see Roediger & McDermott 1995, 810–811). All these elaborations presuppose that rememberers preserve a trace of the original experience. Some researchers think that what participants encode is not individual words but a gist. However, Robins (2016a, 442) argues convincingly that gist-based accounts cannot account for all aspects of the DRM effect. In particular, a pure gist-based account does not explain why the recall of studied words exceeds that of closely associated lures.

5.2 What Is Episodic Memory?

Even the explanation of cases in which experimental participants claim to remember entire events that never occurred occurs against a background that presupposes that actual memories are causally connected, via some actual trace, to some episode in the past. Consider the influential work of Loftus and Pickrell (1995) in which subjects are implanted with a "false memory" that they were lost in a mall during childhood. In attempting to explain why around one-quarter of their experimental participants were induced to falsely remember this incident (with the help of relatives who were in the know and conspired with the experimenters), the researchers presuppose that actual memories are created when an episode leaves a trace in the subject's mind–brain. It is worth quoting them at length:

> The development of the false memory of being lost may evolve first as the mere suggestion of being lost leaves a memory trace in the brain. Even if the information is originally tagged as a suggestion rather than a historic fact, that suggestion can become linked to other knowledge about being lost (stories of others), as time passes and the tag that indicates that being lost in the mall was merely a suggestion slowly deteriorates. The memory of a real event, visiting a mall, becomes confounded with the suggestion that you were once lost in a mall. Finally, when asked whether you were ever lost in a mall, your brain activates images of malls and those of being lost. The resulting memory can even be embellished with snippets from actual events, such as people once seen in a mall. Now you "remember" being lost in a mall as a child. By this mechanism, the memory errors occur because grains of experienced events or imagined events are integrated with inferences and other elaborations that go beyond direct experience. (Loftus & Pickrell 1995, 724)

These psychologists are at pains to explain how such mental states can be generated, and their explanation is parasitic on an explanation of how actual memories are formed. Admittedly, the details of memory formation have not been provided, in particular of how events or episodes generate representations ("grains of experienced events"), which are then stored as traces, and later retrieved. Yet there is a pervasive assumption in much memory research that a process of this kind occurs in the case of genuine memories, even though the cognitive sciences have not yet uncovered the specifics of the hypothesized process, whether at the algorithmic or implementational levels.

The standard explanation of why actual episodic memories succeed (when they do succeed) in revealing details about the past is that they are causally linked in such a way to past events as to enable them to faithfully transmit information. Indeed, cognitive scientists explain memory errors (lure words in the DRM paradigm or "false memories" in Loftus' experiments) against the background of successful memories, which they assume involve traces of past events. The details are still obscure, or at least there is no consensus on them, but something like a causal transmission process is bound to be in place for such an explanation to succeed. As already mentioned, this inference to the best explanation does not require episodic memories to be wholly or even mostly accurate or veridical when it comes to representing the past. We do not even need to assume that memories are veridical more often than not, or that the track record of memory is better than, say, imagination. Even if it turns out that memory is only occasionally accurate or veridical, those occasions would need an explanation. Compare: If dreams enabled us to predict the future, even sporadically, that fact would cry out for an explanation. The explanation of the fact that states of episodic memory succeed in accurately representing past events is that the mental states in question transmit information from that past event more or less faithfully by virtue of being causally connected to it via a trace or representation of the past event. That would seem to be sufficient scientific justification for associating episodic memory with a causal link to a past episode.[13]

We have now encountered the two most prominent approaches to characterizing episodic memory: phenomenal and etiological. Both of these are attested in the philosophical and psychological literatures. Indeed, the lengthy passage from Loftus and Pickrell quoted above can be interpreted as linking the two characteristics, in attempting to explain why certain states of mind that do not have the requisite etiology can yet appear to subjects to be real memories, that is, have the right phenomenal features. There is a large body of empirical work that relies on these two ways of characterizing episodic memory. Given that this is the case, we can ask two questions. First, what reason do we have for thinking that these two purported features of episodic memory (usually) coincide? Second, do either or both of these features give us grounds for thinking that states of episodic memory or the cognitive capacity that generates such states corresponds to a real kind? Before tackling these questions, it is worth looking at some challenges to the notion that episodic memories (states) or episodic memory (capacity) are real kinds.

---

[13] Even though this is not strictly necessary for my argument, one might go further and argue that episodic memory is a reliable system, meaning that it produces mostly true beliefs (see Michaelian 2011a; Michaelian 2011b). Michaelian (2011a, 333) adds a condition to the causal theory: "the causal chain must go not only via a memory trace but through a properly functioning (that is, a reliable) memory system." He thereby replaces the causal theory with a "causal reliabilist theory of memory," though he abandons this theory of memory in later work.

5.3 Empirical Challenges

A number of cognitive scientists have challenged the existence of a distinct category of episodic memory and have instead posited that humans have a more general capacity for "mental time travel" (Suddendorf & Corballis 1997), "episodic hypothetical thinking" (De Brigard 2014), or "constructive episodic simulation" (Addis 2018), into which episodic memory is submerged. Although the connection between episodic memory and certain forms of imaginative, hypothetical, or prospective thinking had been drawn at least as early as Tulving (1985), this recent work goes further by claiming that there is no distinct category of episodic memory at all and that what we call "episodic memory" is really just a manifestation of a cognitive system with a much broader function.[14] The evidence that this hypothesis is based on can be divided broadly into two main categories: neuroimaging evidence and evidence from neuropathology. In this section, I will survey this evidence briefly and try to understand its implications for the question of the status of episodic memory as a real kind.

There is an appreciable body of neuroimaging evidence from fMRI studies indicating that tasks requiring the use of episodic memory and those that involve simulation recruit the same neural network, or at least overlapping neural regions. Addis (2018, 73) claims that the neural system in question is the "default mode network" (DMN), a set of interconnected regions that include "medial temporal lobes, medial aspects of the frontal and parietal cortices, inferior lateral parietal cortex and lateral temporal cortex." The DMN was originally identified as a result of imaging experiments in which participants lying in the scanner between tasks generally exhibited activity in a certain combination of neural regions.[15] Accordingly, it was posited that these regions jointly subserve "default" activity in the brain, that is, they are active when a subject is not carrying out any particular cognitive task. Later research maintained that the network was associated with daydreaming or "mind-wandering," since this is presumably what subjects are doing when they are not engaged in specific cognitive tasks (e.g. reading a word, solving a puzzle). But Addis (2018) and other researchers claim that the DMN performs the function of imagination or simulation, and that this cognitive function subsumes what we

---

[14] Addis (2018, 70) explains her change of position as follows: "... instead of conceptualising imagination as relying on episodic memory, we should consider memory and imagination as but two products of a constructive simulation system ..."

[15] Different researchers sometimes enumerate somewhat different regions as being implicated in the DMN.

have labeled "episodic memory." Other theorists (including Addis herself in earlier work; see e.g. Schacter, Addis, Hassabis, et al. 2012) take a more moderate view, claiming only that this network, or parts of it, are involved in both episodic memory and some types of imaginative thinking, perhaps because imagination draws on memory in some ways.

A number of neuroimaging studies are thought to support the view that there is no distinct capacity for episodic memory, but rather a single cognitive system for episodic recollection, future projection, episodic counterfactual thinking, and similar cognitive processes. These studies tend not to find an exact coincidence between the networks recruited in the tasks relevant to all (or some of) these cognitive functions, but considerable overlap. For instance, one study found "significant neural overlap between brain regions engaged during autobiographical recollection and those engaged during episodic counterfactual thinking" (De Brigard, Addis, Ford, et al. 2013, 2408). Needless to say, significant overlap when it comes to neural correlates cannot be taken as conclusive evidence for the equivalence of the corresponding cognitive capacities, or their subsumption into some overarching cognitive function. There are a number of alternative interpretations for findings of this type. One would be to maintain that episodic memory relies upon some of the same resources as these other types of thinking (i.e. prospective thinking, hypothetical thinking, counterfactual thinking, imagination), though it is a distinct cognitive capacity. This is similar to what was described as Addis' former view above (see e.g. Schacter, Addis, Hassabis, et al. 2012), namely that imagination draws on memory, or that they both depend on common cognitive systems that perform other functions. One candidate proposed in the psychology and neuroscience literature for this function is "scene construction" (see e.g. Suddendorf, Addis, & Corballis 2009; Szpunar 2010). Szpunar (2010, 157) describes this cognitive function as a capacity for constructing a "coherent representation of a specific scenario," which may well be needed both for reconstructing memories *and* for constructing states of imagination or prospective thinking. In other words, activation in this network might correspond to a component cognitive process that is common to these different functions. It need not indicate identity among these cognitive functions. A completely different interpretation of these findings would have it that neural overlap, or even coincidence, as indicated by neuroimaging studies is inconclusive when it comes to the cognitive or computational level because of the phenomenon of neural reuse. As discussed in previous chapters (see especially Section 1.4), the activation of a neural region or indeed an entire network is not a decisive indication of cognitive function.

5.3 Empirical Challenges

Even without taking into account different levels of activation in various regions that emerge across these tasks, neural reuse holds that the very same regions or networks can be deployed for different cognitive tasks depending on phenomena such as neuromodulation of neural circuits by chemical and genetic means. Therefore, the neuroimaging evidence does not indicate decisively that episodic memory is not a distinct cognitive capacity. It can be interpreted as implying either that it taps into some of the same cognitive resources as other cognitive capacities, or that it recruits some of the same neural resources as they do.

Another source of evidence cited in support of the hypothesis that episodic memory is not distinct from other cognitive capacities, particularly those for prospective or future thinking, comes from neuropathology. This evidence dates back to some of Tulving's seminal work on patient KC, who was involved in a motorcycle accident at age thirty that resulted in multiple brain lesions particularly in the medial temporal lobes, and suffered from severe amnesia as a result. He experienced both anterograde and retrograde amnesia for episodic memories, meaning that he could not recall personally experienced events that occurred before the accident, nor could he form new memories of events that he experienced after the accident (Tulving 1985, 4–5; Tulving 2002, 13–14). In addition to these deficits, KC was unable to engage in thinking about the future, and when given the opportunity to do so, would generally draw a blank (Tulving 1985, 4). Subsequent research on some other amnesic patients has revealed similar deficits when it comes to prospective thinking. In particular, some other patients with damage in the medial temporal lobe (which includes the hippocampus and is often considered to be an integral region in the DMN) also exhibit similar deficits. One of the most widely cited studies tested a group of five amnesic patients with bilateral hippocampal damage to determine how they compared to controls in imagining future experiences. The study found that such patients not only suffered from severe amnesia but had serious deficits when it came to imagining new experiences and fictitious scenarios (Hassabis, Kumaran, Vann, et al. 2007). Such findings are sometimes cited to support the contention that episodic memory is not a distinct cognitive capacity, but rather is just a manifestation of a broader cognitive capacity that enables us to engage in "mental time travel," whether into the past or the future. However, this study posits not that episodic memory is one manifestation of a cognitive capacity that is also dedicated to imagination and prospective thinking, but rather that these capacities draw on some of the same cognitive resources. In particular, the researchers conclude that the "patients' imagined experiences were strikingly deficient in spatial coherence, resulting in their constructions being fragmented and lacking in richness" (Hassabis, Kumaran, Vann, et al. 2007, 1729). Hence, they hypothesize that the hippocampus, which was the site of neural damage in all five participants, "may make a critical contribution to the creation of new experiences by providing the spatial context or environmental setting into which details are bound ..." (Hassabis, Kumaran, Vann, et al. 2007, 1729). Therefore, the results from neuropathology do not seem decisive in rejecting the reality of a capacity of episodic memory. Like some of the neuroimaging studies mentioned, one interpretation of these findings would simply be that episodic memory and the ability to describe future scenarios draw on some of the same cognitive resources, and that the patients in question are impaired in these other abilities, which affect episodic memory as well as other cognitive capacities. Finally, it is worth mentioning that patients with severe deficits in episodic memory do not seem to be impaired when it comes to another aspect of future thinking, namely temporal discounting. When some amnesic participants are asked whether they would prefer to receive a monetary reward in the present (e.g. $50 now) or a somewhat more valuable monetary reward in the future (e.g. $100 in three years), they respond similarly to controls, namely by discounting future rewards to roughly the same degree (Kwan, Craver, Green, et al. 2015). Hence, it is not the case that amnesic patients are thoroughly compromised in their ability to think about the future.

We have now encountered two prominent challenges to the reality or kindhood of episodic memory as a cognitive capacity, which argue for the elimination of episodic memory, but we have also seen that neither of these eliminativist proposals is conclusive. There are other, more limited, challenges both to episodic memory as a type of *capacity* and to episodic memories as types of *states*, which tend to go by the name of "constructivism." It is difficult to summarize the variety of positions that are commonly tagged with this label, but perhaps the single common denominator is that episodic memory is not a preservative capacity that simply encodes, stores, and retrieves traces of past experiences. Rather, episodic memories are composed of representations that are constructed in response to endogenous or exogenous retrieval cues and combine resources from a variety of sources. These representations may harbor traces of past experiences, but they are cobbled together using information from semantic memory, perception, the retrieval cue, and elsewhere. As I have just characterized this family of positions, it does not obviously challenge the kindhood of episodic memory. But a more radical variation of this view, which denies the existence of traces altogether, or denies that they are a necessary component of memory states, does challenge the kind as I have characterized it so far. In the previous section, it was proposed that episodic memory states are distinguished at least partly on the basis of their etiology. That etiology is a matter of harboring a trace of a past experience, so if traces are denied altogether, that would undermine the proposed characterization of this representational kind. Moreover, it would erase the distinction between this representational kind and other such kinds, like states of imagination, which need not bear such traces. In practice, few cognitive scientists adopt such a radical position and those who do tend to subscribe to the eliminativist position just discussed, which would submerge episodic memory into a broader cognitive capacity. Most of those who consider themselves constructivists still incorporate a trace into their memory states. Hence, the etiology of memory is still important to most brands of constructivism, even though it is not emphasized to the exclusion of other aspects of the episodic memory state. For example, Schacter (1996, 71) puts it as follows:

> When we remember, we complete a pattern with the best match available in memory; we do not shine a spotlight on a stored picture. The idea that a memory is an emergent property of the [retrieval] cue and the engram is difficult to accept. We must leave behind our familiar preconceptions if we are to understand how we convert the fragmentary remains of experience into the autobiographical narratives that endure over time and constitute the stories of our lives.[16]

The reference to the "engram," which is the hypothesized neural substrate of the memory trace, as well as the "fragmentary remains of experience," indicate that Schacter is committed to the idea that states of episodic memory incorporate some distinct representational resources caused by a past experience. This is fairly typical of constructivist approaches to episodic memory, even though few if any cognitive scientists would claim to know how precisely the capacity of episodic memory concocts memory states out of a variety of representational resources, nor how these resources combine to issue in the memory state itself. But the existence of a trace, as at least one component of states of memory, is a presupposition of many if not most constructivist approaches to episodic memory.

In this section, I have considered some of the most prominent challenges to the idea that episodic memory is a real cognitive kind, both as a

---

[16] See also Tulving (2007, 67): "... the memory trace is not just mere residue, or after-effect of a past experience, not just an incomplete record of what was. It is also a recipe, or a prescription, for the future."

type of cognitive capacity and as a type of mental state. In the case of the eliminativist position that contends that there is no capacity of episodic memory distinct from a broader capacity for simulation, I argued that the evidence from neuroimaging and neuropathology is inadequate to support the position. As for constructivist approaches to memory, many of them are still committed to the existence of a memory trace. Memory traces are seldom explicitly defined in the empirical literature, but one notable exception is due to Tulving (2007, 66): "A memory trace is the neural change that accompanies a mental experience at one time (time 1) whose retention, modified or otherwise, allows the individual later (at time 2) to have mental experiences of the kind that would not have been possible in the absence of the trace." This definition would seem to accord broadly with the notion of a memory trace utilized by Martin and Deutscher (1966), which posits an uninterrupted and nondeviant causal chain between a past experience and a representation at the point of recall, though Tulving explicitly identifies the trace with a neural change, while Martin and Deutscher do not. Moreover, Martin and Deutscher describe it as a "structural analogue" of the original experience, but this additional condition seems unnecessary (cf. Robins 2016c). What seems to be required is that it be caused by the past episode, represent it, and be recoverable as pertaining specifically to it (in a word, *traceable* back to that episode). In Section 5.5, I will argue for a distinction between the *trace*, which can be identified at the computational level, and the *engram*, its supposed neural substrate, which belongs to the implementational level. But before doing so, I will make the case in Section 5.4 that episodic memory (both the state and the capacity) is a real cognitive kind.

5.4 Episodic Memory as a Cognitive Kind

In previous sections, I introduced etiological and phenomenological approaches to identifying episodic memory, and I also outlined some challenges to episodic memory being a real kind, at least if it is understood to have a distinctive etiology and phenomenology. I argued that these challenges can be resisted on the basis of available empirical evidence, but that is not enough to establish episodic memory as a real cognitive kind, either as a kind of mental state or capacity. In order to do so, we need to determine whether it can be identified with a set of causal properties. In particular, we will need to specify these properties more precisely. Then, we will have to ascertain how these properties relate to one another, if at all. Finally, we will need to determine whether episodic memory states or

5.4 Episodic Memory as a Cognitive Kind

the capacity of episodic memory has the causal profile to constitute a real cognitive kind. In this section, I will attempt to tackle these tasks, focusing primarily on episodic memory states.

When it comes to the posited phenomenal property of episodic memory states, which Tulving characterized as "autonoetic," various researchers have tried to characterize it in qualitative terms. As we have seen, participants in experimental conditions are probed for their episodic memory using the contrast between the verbs "remember" and "know." Moreover, many researchers have described episodic memory states as characterized by the quality of reliving an original experience or by an awareness that the event occurred in one's past. Klein (2015, 2) says that an episodic memory state "includes the feeling that one is reliving a past experience – that is, it provides a directly-given, non-inferential sense that one's current mental state reflects a happening from one's past." Similar proposals have been made by numerous others. However, on most psychological and philosophical accounts, the phenomenal properties of episodic memory are not causally inert or epiphenomenal, but have a distinct functional profile that causally influences both behavior and verbal reports (cf. Boyle 2020). Moreover, in healthy human subjects, episodic memories are thought to be readily recognizable as such and can be distinguished from semantic memories, as well as from mental states such as perceptions, dreams, and imaginations.

As for the purported etiological property of episodic memory states, there are two prominent difficulties pertaining to it, one of which is epistemic and the other ontological. On the epistemic front, there is a general problem when it comes to the reconstruction of etiology, since delineating the causal chain that links a unique past event to a current mental state is typically not amenable to direct investigation, and must be inferred indirectly, whether in everyday or experimental settings.[17] As for the ontological challenges, these have to do with the now widely acknowledged constructive nature of states of episodic memory. As we saw briefly in Section 5.3, constructivists about episodic memory hold, on the basis of a considerable body of empirical evidence amassed over several decades, that episodic memory states are generated from memory traces as well as other representational resources. Thus, such states are widely thought to be hybrids, wherein traces may be amalgamated with non-trace

---

[17] This obstacle might be overcome using recently developed experimental techniques in optogenetics that attempt to localize memory engrams (Robins 2016b; Liu, Ramirez & Tonegawa 2014). But, as I will try to argue in Section 5.5, we should not presuppose a direct correspondence between the trace and the engram.

representational content or otherwise incorporated with representational resources that do not derive from the event that is the supposed subject of the memory. This means that the posited trace may not be capable of being isolated or distilled from the mental representation that is the state of episodic memory, since it is ontologically entangled with it in certain ways. Moreover, this will inevitably lead to questions about the extent to which the trace needs to figure in the resultant representation for it to qualify as a genuine state of episodic memory. I cannot pretend to have answers to these questions, but will try to say something about each of them in order to show that the etiological account of episodic memory is not hopeless.

The epistemic challenge of ascertaining whether a current representation incorporates an etiological trace is not insurmountable. Especially under experimental conditions, researchers can control aspects of an event or stimulus that participants are asked to remember and can compare it to their responses at a later time. As mentioned in Section 5.2, when subjects correctly recall previously shown words or objects, it is possible that what they say they remember are merely lucky guesses, but with enough trials and sufficient numbers of participants, experimenters can rule out guesswork and can determine when participants have actually remembered, and infer that they have indeed retained a trace of a past event. Numerous experimental paradigms have been designed to determine when subjects retain traces of past events.[18]

The ontological question has to do with the extent to which "impurities" can be tolerated in an episodic memory state. Memory researchers have detailed numerous types of distortions that can affect episodic memory and it is now well known that a great many of our memories are subject to these systematic errors (for an overview, see Michaelian 2011). This means that memory states are not simply preserved mental representations of past experiences, but incorporate much else besides. Should we consider a mental state to be an episodic memory if it contains a negligible representational trace, despite the fact that it is largely composed of other types of

---

[18] As already mentioned in Section 5.2, a standard distinction made in the empirical literature to mark this phenomenological difference is that between "remembering" and "knowing" (e.g. Tulving 1985). This has been elaborated in instructions for experimental participants in a large body of research. For example, one much-used set of instructions (Rajaram 1993, 102), reads in part: "If your recognition of the word is accompanied by a conscious recollection of its prior occurrence in the study list, then write 'R.' 'Remember' is the ability to become consciously aware again of some aspect or aspects of what happened or what was experienced at the time the word was presented (e.g., aspects of the physical appearance of the word, or of something that happened in the room, or of what you were thinking and doing at the time) … 'Know' responses should be made when you recognize that the word was in the study list but you cannot consciously recollect anything about its actual occurrence or what happened or what was experienced at the time of its occurrence."

representation?[19] This problem can be finessed by bearing in mind that the *capacity* of episodic memory is ontologically prior to the *states* of episodic memory, since the capacity is causally responsible for generating the states. That means that episodic memory states can be individuated primarily by their having been produced by the capacity of episodic memory. This dispenses with the need for trying to say how much of a trace is enough for a state to be a memory. If, as I will go on to argue in this section, there are grounds for positing a *capacity* of episodic memory whose function is in part to transmit traces of past experiences, then the states generated by such a capacity will generally incorporate traces. Depending on how exactly the capacity works and its propensity to malfunction, it may sometimes generate states that do not have such traces or have negligible traces, but those states can still be considered states of episodic memory so long as they are outputs of the episodic memory system.[20]

If the epistemic and ontological challenges to the proposed features of episodic memory can be met, it remains to be seen what we can say about the causal profile of episodic memory and specifically how its etiological and phenomenal features might fit together. There is no shortage of introspective and experimental evidence to suggest that mental states that have the phenomenal properties typical of episodic memory states do not always have a nondeviant causal link to specific episodes in the lives of subjects,[21] and conversely (although less prominently), that some mental states that do have such a causal link do not have the right phenomenology. But despite the fact that these two features do not coincide invariably, it seems frankly surprising that they coincide so frequently and reliably. As we have seen, even work that emphasizes the susceptibility of subjects to "false memories"

---

[19] Moreover, if some mental state incorporates traces from a number of past events, $E_1, ..., E_n$, as well as other representational content, what should we consider it to be a memory of, $E_1, ..., $ or E_n, or all of the events, or none of them? There does not seem to be a clear answer to such questions and we may have to live with some vagueness when it comes to saying what some memories are memories of.

[20] It is consistent with this account that episodic memories start out being largely traceable to their sources but degrade substantially over time to the point that they no longer incorporate traces. What is crucial is that the relevant states are products of the posited episodic memory capacity, not that they incorporate traces. Thanks to an anonymous referee for urging that I clarify this point.

[21] But note that states that have the right phenomenology without the requisite etiology can often be explained by observing that they can be traced back to similar past events. For example, here is how Roediger and McDermott (1995, 811) try to explain why participants claim to "remember" (not just to "know") that a word that wasn't on the original list was on the list, that is that they have the phenomenological experience of having studied the word: "The enhanced remember responses may be due to subjects' actually remembering the experience of recalling the item, rather than studying it, and confusing the source of their remembrance; similarly, it could be that subjects remember thinking about the item during the study phase and confuse this with having heard it." See also the explanation of Loftus and Pickrell (1995), cited in Section 5.2.

acknowledges this fact. What is in need of explanation is the fact that the mental states that have the requisite phenomenal quality are reliably traceable to the actual past events that they purport to represent, and vice versa. Without denying our susceptibility to the implantation of "false memories" and the various memory distortions that we may fall victim to, the coincidence of these two seemingly unrelated features is in need of explanation. One way of accounting for this coincidence is by positing the existence of a cognitive system or capacity that both transmits traces of particular past events and produces representations with certain phenomenal properties. Hence, the existence of a kind of mental state that possesses these properties can be tied to the existence of a mental capacity that tends to generate states with both types of properties, etiological and phenomenological. States of episodic memory can be hypothesized to be generated by a type of capacity whose function is to transmit information about the past, though its function may not be to do so perfectly or even with high fidelity. The resultant states' being traceable to past experiences allows them to carry information about those experiences. To be sure, it does not *guarantee* that they will carry information about past episodes, but to the extent that they do, this is a result of their etiology, and is explainable by it (as argued in Section 5.3). This capacity of episodic memory states to carry information about the past, albeit modified in the various ways stressed by constructivists about memory, may have a variety of different functions. There are numerous theories currently in play about the function of the capacity of episodic memory, many of which tie its function to the ability of cognizers to deal with future eventualities and contingencies.[22] If that is the case, then mere preservation or archiving is not the point, which would go some way toward explaining some of the results stressed by constructivists about memory.

Consider a theory about the function of episodic memory that is due to Klein, Cosmides, Tooby, et al. (2002), who propose an evolutionary account of this cognitive capacity. On their view, episodic memory functions, in part, to restrict the scope of generalizations provided by semantic memory. In other words, part of the point of episodic memory is to provide exceptions to the generalized information supplied by semantic memory. They argue that in addition to a capacity whose function it is to

---

[22] One way of thinking about the function of episodic memory is to ask what it contributes that semantic memory does not. Boyer (2009, 4) poses a question as to the evolutionary point of episodic memory, noting that semantic memory gives humans "stable, declarative, and accessible knowledge of the environment," which would seem to be sufficient from an evolutionary point of view. By contrast, the point of episodic memory, which provides "information about unique, specific situations … encountered in the past," is less obvious (Boyer 2009, 4).

deliver generic information gleaned from the past (semantic memory), we need a capacity that supplies exceptions in the form of unique episodes (episodic memory), since both are valuable when it comes to planning and decision-making, particularly in our social interactions with others. As Klein, Cosmides, Tooby, et al. (2002, 318) elaborate: "A generalization is most useful when its scope is delimited: when it is accompanied by information specifying those situations in which it does not apply." This quick summary oversimplifies what these researchers say about the function of episodic memory. But my aim is not to argue for the view as much as to consider it as a candidate theory that would attribute a plausible adaptive function to the capacity of episodic memory. If it can be shown that episodic memory has one or more adaptive functions that set it apart as a distinct capacity, that would help establish it as a cognitive kind. That is not the only way to establish episodic memory as a cognitive kind, since there are surely such kinds that are not adaptive.[23] But it would put this capacity on a par with other cognitive systems, whose causal-functional profile establish them as real kinds. Moreover, the capacity itself would be an etiological kind that has evolved for a specific purpose, namely to generate cognitive states with certain characteristics. In this case, the chief characteristic of such states would be to bear information about past events that register exceptions to generic facts, notably about our social partners. If this hypothesis is correct, these information-bearing states are causally efficacious in equipping thinkers to plan and make decisions in the social domain, armed not just with generic information but more specific episodic information as well. This account also goes some way toward accounting for a moderate constructivism about episodic memory, since representations that enable us to plan and make decisions are likely to be ones that combine specific information about past episodes with generic information in such a way as to enable us to navigate our environment, particularly social aspects of the environment. Finally, this type of account forges a link between the capacity being a real kind and the states generated by the capacity being members of a real kind (see Figure 5.2).

Although this account is somewhat speculative, it is illustrative of an account that would make episodic memory a real cognitive kind. Not only does the capacity of episodic memory constitute an evolved adaptive capacity on this account, the states that it generates also belong to a real kind. That is because they not only share an etiology but they also play a causal

---

[23] Robins and Schulz (forthcoming) argue that the capacity for episodic memory is an evolutionary by-product rather than an adaptation.

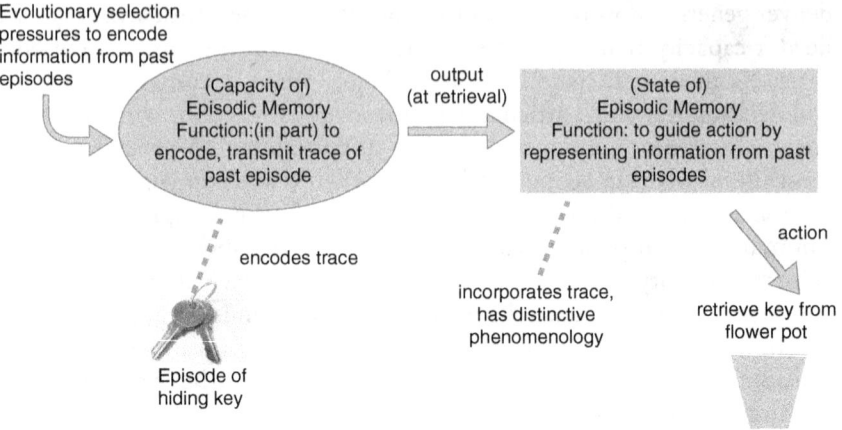

Figure 5.2. Schematic diagram of *episodic memory* (capacity) and *episodic memory* (state) showing their relationship and their partial causal and etiological profiles.

role in the cognitive life of thinkers who are able to have such states, namely in facilitating social decision-making. In addition, this account can also be used to shed light on the phenomenal properties of episodic memory states. Why is it that such states share (by and large) a certain phenomenology, and what is the relationship between their etiology and their phenomenology?[24] One prevalent theory in the psychological literature is that the phenomenal properties of episodic memories enable the thinker to distinguish episodic memories, which bear information about the past, from other mental states, such as states of imagination. On this "source-monitoring" or "reality-monitoring" framework, episodic memories have a distinctive phenomenology to set them apart from mental states that have a different source (e.g. imagination, inference) and are not rooted in a specific past event. For example, Szpunar (2010, 148) writes: "According to reality-monitoring theory ..., these differences in phenomenological characteristics [between representations of imagined and actual events] are essential because they play a pivotal role in helping people to discriminate imagined events from actual memories." Indeed, even Addis (2018, 68) acknowledges that there is a wealth of empirical research indicating phenomenal differences between episodic memories and states of imagination: "Relative to imagined representations, memories typically have more sensory detail ..., contextual

---

[24] Compare Hoerl (2001, 324): "There must be a connection between the specific epistemological status of episodic memory, as a particular way of retaining knowledge of events we have experienced in the past, and its phenomenology."

clarity ..., and consistency with pre-existing knowledge that makes them more plausible ..." This shows how the etiology and phenomenology of episodic memory may be linked and why these causal features go hand in hand. To summarize, on this account, the capacity of episodic memory is an evolved system whose function is (at least in part) to transmit representations of past episodes that constitute exceptions to generic information gleaned from the past (information that is stored in semantic memory). The representational states generated by the capacity of episodic memory are characterized primarily by their etiology (the link to past episodes), but they also play a synchronic causal role in cognition, in that they enable the thinker to plan and make decisions about the future, equipped with specific information about the past. Moreover, these states are also typically characterized by a distinctive phenomenology to enable the thinker to distinguish such states from those of imagination, inference, and indeed states of semantic memory, which do not have such a phenomenology.

This may seem like an adaptive "just-so story," yet any complex cognitive capacity that delivers distinctive states and is doubly dissociated from other systems, is likely to be an evolved adaptive capacity. Other functions have been proposed for episodic memory, such as motivating prosocial behaviors based on past experiences (Boyer 2009), anticipating future regret in order to plan accordingly (Hoerl & McCormack 2016), or demarcating epistemic authority for the purpose of communicating with others (Mahr & Csibra 2018), and indeed, episodic memory may have evolved to serve multiple functions. But whatever function or functions it has been selected for, the capacity of episodic memory can be argued to be a real kind both due to its etiology (hypothesized history of natural selection) as well as for the causal role that it plays in the lives of the thinkers who have this capacity, namely to fulfill the function for which it was selected, in addition to possibly fulfilling other causal roles. Like many other adaptive capacities, episodic memory may have come to play other functional roles in the lives of thinkers, many of which have been investigated by social psychologists, such as: to enhance self-understanding, reduce boredom, harbor grudges, reminisce about dead loved ones, cope with thoughts of mortality, teach lessons to young people, and forge ties with friends, to name some (see e.g. Harris, Rasmussen, & Berntsen 2014). Moreover, the states that are the outputs of this capacity are also members of a real kind by virtue of sharing an etiology (being caused by past experiences and generated by the capacity of episodic memory) and having certain adaptive causal features, including phenomenal properties. The claim that episodic memory is a real cognitive kind does not depend on the specific adaptive hypothesis described above, but is compatible with any number of hypotheses about its adaptive function and its current causal role in the mental lives of thinkers. The same goes for the states that are the outputs of this system. This particular hypothesis not only posits a link between the evolved function of the capacity of episodic memory and the causal role it plays for cognizers, it also bridges the gap between this role and the phenomenal properties of the states produced by the capacity. On this account, the capacity is not assumed to be foolproof. As we have seen, there is experimental (and anecdotal) evidence to suggest that states with the requisite phenomenal properties may not bear information about episodes in one's personal past. Such missteps may arise along the lines that Loftus and Pickrell (1995) suggest, as mentioned in Section 5.3, or for some other reason. Cognitive systems are not flawless and may often malfunction, so this finding should not be surprising. But this account at least explains the convergence between etiology and phenomenology, whereby those states that are causally linked to past episodes in the life of the thinker and represent them are also phenomenally distinctive.

5.5 Is Episodic Memory a Neural Kind?

As argued in previous sections, there is a widespread view in the cognitive sciences, as well as in philosophy, that the presence and persistence of a trace is a necessary element in a state of episodic memory (cf. Robins 2017, 76). Some researchers in the cognitive sciences prefer to speak simply in terms of "persistence" rather than a trace. One researcher puts it starkly as follows: "Persistence is an essential property of memory. Indeed if memories did not persist at all, there would be no memories" (Thompson 2007, 199; cf. Eichenbaum 2007; Lisman 2007). A commonly mentioned obstacle to the scientific study of traces or persistence in memory research is that behavioral measures are incapable of distinguishing cases in which a memory trace is absent from those in which it persists but there is a failure in retrieval. Partly for this reason, some researchers propose that traces should be investigated directly by examining their neural underpinnings. Lisman (2007, 206) puts it this way: "My view is that once memory molecules responsible for persistence are identified, the investigation of persistence can be put on a solid footing. It will be possible to separate persistence from retrieval processes by direct biochemical tests."[25] It is tempting to

---

[25] The reference to "memory molecules" may not be hyperbole. At least since the 1960s, some researchers have posited a role for RNA molecules in encoding and storing memories (see e.g. Bédécarrats & Glanzman 2018).

5.5 Is Episodic Memory a Neural Kind?

stake the claim that episodic memory is a real kind on certain neuroscientific discoveries, making it conditional on discovering a neural correlate of memory traces. As we have seen, in many scientific discussions of memory, the engram is invoked as the neural correlate of the trace that is posited to make an appearance in all episodic memories. Although I would not rule out the discovery of the elusive engram, in this section I will argue that the claim that episodic memory is a real kind should not be conditional on discovering a neural correlate for it.

Despite the fact that we may occasionally be unable to distinguish failure of persistence from failure of retrieval, it is clear that there are many instances in which we can, and that we have robust methods of doing so. Take, for instance, a case from classical conditioning rather than episodic memory. After a conditional response has been rendered extinct, rapid relearning of the response generally demonstrates persistence (cf. Thompson 2007, 199–200). This is often used to show that the conditional response persisted but that it ceased to be retrieved after a certain point. Similarly, in episodic memory, when certain retrieval cues succeed in generating the memory after other cues have failed, we can infer that their failure was not due to lack of persistence or the absence of a trace. At most, it can be said that in some cases of failure of memory we may not be able to distinguish failure of persistence from failure of retrieval. But that would seem to be a feature of any attempt at reconstructing etiology. (Compare: In some instances, we cannot distinguish cases of phenotypic homology from ones of convergent evolution.)

Still, it may be said that the proof is in the pudding, and in this case, that means locating the actual trace that has persisted, or indeed its neural correlate, the engram. In response, it is worth making a distinction between the engram and the trace. As I have tried to explicate it, a trace is a theoretical posit that explains the success of episodic memory.[26] It is a cognitive entity at what Marr calls the computational level that is causally linked to a past event and carries information about that event. A trace is posited because of the need to explain *how* it is that memory states succeed (when they do succeed) in capturing aspects of past events and enabling cognizers to meet present challenges. Indeed, traces are also invoked to explain some cases of misremembering, as Robins (2016a) has argued (see Section 5.2). Traces feature in computational-level explanations that relate

---

[26] See Rosen (1975) for an early argument to the effect that memory traces are theoretical posits that perform an essential explanatory role; for a critique of that argument, see Heil (1978). Thanks to Amy MacKinnon for first raising the possibility in conversation that the trace is different from the engram.

cognizers to their environments and their causal history, both ontogenetic and phylogenetic. They also enable us to explain *why* cognizers behave as they do, *why* they are sometimes successful in doing so, and *why* such behavior may be the result of certain cognitive adaptations. By positing a trace, one makes no commitment to the nature of the neural structures that might implement such traces, though in humans and similar creatures something at the neural level must ultimately contribute to the encoding, persistence, and retrieval of traces. But traces may be multiply realized relative to their neural correlates, when it comes to different episodic memories in the same individual, across individuals, and across species. They may also be highly distributed rather than localized, and indeed, there is considerable evidence to suggest that episodic memory traces cannot be located in a specific brain region (see e.g. Thompson 1991; Nadel, Samsonovich, Ryan, et al. 2000). Moreover, although the trace can be distilled in theory and in certain experimental settings from other ingredients of an episodic memory (e.g. information from semantic memory, perceptual cues, and other sources), it is not clear that the neural correlate of the trace should be thought of as a distinct or separable neural component of an occurrent neural state. Finally, it should also be emphasized that the trace may be dispositional in nature, which means that its neural correlate need not be manifested when the memory is not being retrieved. If so, the neural correlate of the trace as manifested in an occurrent mental state is not likely to be a neural structure or process that can be identified in the brain when the memory is not actually being retrieved.

If any of these features hold of episodic memory traces (multiple realization, distribution, dispositionality), a trace cannot be straightforwardly identified with an engram, at least if an engram is thought of as a standing neural structure, set of synaptic strengths, or pattern of neural activation. In addition, as I have characterized them, memory traces at the computational level are individuated with reference to their etiology. Two intrinsically identical representations may not both be traces, since they may differ in their causal histories. This means that any attempt to identify traces with neural correlates will have to contend with the fact that traces are individuated extrinsically rather than intrinsically. This poses certain obstacles to investigation of the purported neural correlates of memory traces, at least using current methods. Experimental participants can undergo neuroimaging scans when they are in the process of retrieving a memory, but determining whether the neural state that they are in bears a trace of a past event will be a challenging task given current experimental techniques. These obstacles may not just be practical ones, since as I have argued in previous chapters, etiological individuation is not widely attested in neuroscience and when it is, the etiological factors that are tracked are not always those of psychological or cognitive interest. Various philosophers of neuroscience have emphasized the importance of mechanisms and mechanistic explanation to the neuroscientific enterprise (e.g. Craver 2007; Craver 2013), and to the extent that they are right, this does not comport well with the etiological or functional inquiries associated with episodic memory. Mechanistic explanations generally focus on organized systems and their parts, and explain their activities with reference to spatially arranged components that are contiguous to each other and interact by direct physical contact. Functional explanations, on the other hand, typically involve systems that are spatially dispersed and interact indirectly through intermediaries. As mentioned in Chapter 1, as understood in this book, computational explanations are a species of functional explanation. Computational explanations invoke information-bearing states and processes, which typically transmit information over relatively large spatiotemporal scales. When we say that a trace explains *why* episodic memory succeeds in representing past events, or that the success of memory explains *why* a capacity of episodic memory was selected for in our evolutionary history, we are abstracting away from details of implementation. Of course, there is room also for explaining *how* memory is able to fulfill its functions, and this is often a demand for a mechanistic explanation. But the two types of explanation are somewhat independent of one another, and the scientific research programs that are primarily engaged in one type of explanation have different methods and taxonomies than the ones that engage in the other. Moreover, they pertain to autonomous domains or causal systems, since computational-functional systems are multiply realizable by mechanistic systems.

An analogy from biology might help here. Homologous phenotypic features are generally individuated etiologically and they are identified on the basis of their relations to ancestral characters. Like the attempt to identify genuine memories, our ability to trace homologous features in different lineages to their ancestral antecedents are fallible, but the fossil record sometimes enables us to make such identifications with a high degree of reliability. It might be thought that the ultimate test of a hypothesis concerning homology would be uncovering the genetic basis of the phenotypic feature. But there is ample evidence to show that some homologies persist despite a change in their genetic bases; they come to be sustained by different genetic mechanisms because they perform an adaptive function that responds to certain abiding selection pressures (Brigandt & Griffiths

2007, 634; Kendig 2015). Though investigation of the genotype may inform research into homologous phenotypic characters, it is not decisive and cannot be taken to settle the question of homology. The comparison can be extended further: It is usually impracticable to investigate ancestral genetic material and one can rarely track genotypes historically. As in the neural case, longitudinal investigation of the underlying mechanisms is generally not available to scientists.

Previous sections point to the conclusion that the primary object of investigation in the cognitive sciences is the capacity of episodic memory and its outputs, that is the states of mind that coincide with recall or retrieval, as opposed to the posited stored trace, which seems much less amenable to investigation, let alone the hypothesized neural engram, which I have argued need not coincide with the trace. This makes memory primarily a "retrieval phenomenon," as some cognitive scientists have proposed (Mahr & Csibra 2018, 51; cf. Klein 2013). Episodic memory states are causally efficacious in generating the behavior of cognitive agents and the success of such states in producing appropriate actions is explained by the information-bearing trace. Moreover, the emphasis on retrieval agrees with the moderate constructivist position and the source-monitoring framework, which were discussed in previous sections (see e.g. Koriat & Goldsmith 1996). Episodic memory states need not represent the past with total accuracy as long as they represent specific content that qualifies more generic information, to assist in planning and decision-making.

I have also argued in previous sections that episodic memory states are likely the outputs of a dedicated cognitive capacity or system, the episodic memory system, whose function it is to transmit certain types of information about the past with a distinctive phenomenology. This raises the question as to whether the system as a whole may have a neural correlate, or may indeed be identified with a specific neural network. We already saw in Section 5.3 that some researchers have suggested a great deal of overlap between the activity of episodic memory and neural activation in the default mode network (DMN). It may well turn out that the episodic memory system is subserved entirely by a particular brain network, either the DMN or some other (presumably, partially overlapping) set of interconnected regions, which is responsible for the encoding, storage, and retrieval of representations about past events for use in the present. But before jumping to the conclusion that the capacity of episodic memory can be identified with this or any other neural network, a couple of caveats need to be kept in mind. One is that the DMN has also been identified with other functions, as eliminativists about episodic memory have noted. This means that there may be a high degree of neural reuse when it comes to the network as a whole or its component regions, so an identification of the cognitive system with a neural network is unwarranted. Moreover, if neural reuse is a basic principle of brain organization, we should not expect cognitive systems to be identical with neural regions or networks, but perhaps with those regions or networks in particular configurations, modulated in certain ways, or as they undergo certain levels or patterns of activation. Another, related, point is that some researchers hypothesize that the capacity for episodic memory is a "uniquely human property of the human mind" (Tulving 1985, 1), and if they are right about that, then evolutionary processes might not have had time to effect structural changes in the brain to accommodate this cognitive system. Moreover, if the episodic memory system is an evolved capacity, then it is individuated in part by the adaptive role it plays for creatures endowed with that capacity. That means that if we were to find similar capacities in other creatures, we might not consider them to be instances of episodic memory unless they were the outcome of the same or similar selection pressures. Hence, the capacity of episodic memory may itself be individuated etiologically and any attempt to identify it with its neural substratum will run into the problem that neural mechanisms are usually not so individuated, as already elaborated for the memory trace.

5.6 Conclusion

In this chapter, I have made the case for considering episodic memory (both the capacity and the states it produces) as a cognitive kind. In doing so, I have tried to synthesize philosophical work on memory with work in cognitive psychology. States of episodic memory are widely regarded by memory researchers as being connected via a trace of some kind to a past event or episode in the thinker's life. I argued that this claim is not just a feature of the "causal theory of memory" favored by some philosophers but is in fact implicit in much of the empirical work in cognitive science, and is indeed required for the explanatory work that memory does in both ordinary situations and experimental settings. This means that states of episodic memory are individuated primarily with reference to their etiology or causal history: They can be traced back to past episodes. They are also characterized by a distinctive "autonoetic" phenomenology, which differentiates them from other mental states. I also argued that there is evidence to suggest that humans have a dedicated faculty of episodic memory whose function it is to produce such states. There are a number of theories that purport to explain why such a capacity would be adaptive and I selected one such theory as being illustrative of a range of similar theories. Since these theories of episodic memory consider it to be an adaptive capacity, it is also etiologically individuated with reference to its evolutionary history, as are other adaptive capacities. Moreover, this means that the states produced by such a capacity are doubly etiological, since they are individuated by virtue of being traceable to past events, but they are also individuated in virtue of being outputs of the capacity of episodic memory. Meanwhile, the capacity of episodic memory has the function of producing states that enable thinkers to plan and make decisions based in part on information from the past. Thus, this capacity has certain synchronic causal powers. There are also grounds for thinking that this capacity endows states of episodic memory with a distinctive phenomenology, which explains why states of episodic memory are also (typically) characterized by this phenomenal quality.

This account of episodic memory regards the *capacity* or system as primary since it is responsible for generating *states* of episodic memory, though both can be considered real kinds. It should be clear from this characterization that the cognitive kind *episodic memory* is externalistically individuated in multiple ways, both phylogenetically (in the case of the capacity) and ontogenetically (in the case of the state).[27] This is yet another instance of a cognitive kind whose identity conditions are tied up with etiology and environmental context, and I would argue that this is reflected in the taxonomic practices of many of the research programs that study memory. This mode of individuation is evident in the context of computational inquiries, which attempt to understand the adaptive function of episodic memory and how it enables thinkers to plan and make decisions in their social and physical environments. It is not prevalent in neuroscientific inquiries and taxonomic categories, which are primarily interested in mechanistic causal processes that are spatially circumscribed and take place on a more limited temporal scale.

Although this account of memory is supposed to be responsive both to a range of scientific research programs as well as the vernacular concept of memory, one aspect of the vernacular concept of MEMORY and the verb "remember" that I have not tried to address in this account is its allegedly factive nature. In English, "remember" often takes a that-clause (e.g. "I remember that I ate pizza for breakfast last Sunday"), and assertions involving this verb presuppose that the content of the that-clause is true. At least on one view of the truth-conditions of such assertions, if the proposition

---

[27] Debus (2016) has also argued for a kind of externalism when it comes to episodic memory states, but she holds that past events are partly constitutive of mental states of remembering.

5.6 Conclusion

expressed by the that-clause is false, the assertion is neither true nor false and it cannot be asserted. Like "know," "recognize," "realize," "learn," and "report," "remember" is supposedly only correctly applied when it is said of something true. On the theory of episodic memory presented here, factivity cannot be upheld simply because nothing about the account guarantees that episodic memories will always be true. Traces represent the past, but (as constructivists emphasize) they may be so entangled with other, non-veridical information, that the resulting content of the episodic memory may be false. Some of the results of the experiments reported in Loftus and Pickrell (1995) may be good examples of this: The participants' mental states may incorporate traces of past experiences as well as traces of suggestions planted by the experimenters, but the ensuing memories are false. If we are interested in giving a naturalist account of the capacity of episodic memory and the states that it generates, we cannot also insist that episodic memory states are true by definition (as factivity would have it). Nothing guarantees that a cognitive capacity that forms part of our psychological endowment will only generate truths, and a definitional insistence on this is bound to collide with empirical results from cognitive science. In addition, it is worth noting that, like the verbs "report" and "learn," some uses of "remember" in English are not factive. The sentence, "The police reported that the culprit had fled," has a factive reading, on which what the police said is true, but it also has a non-factive reading, according to which it need not be. Similarly, it seems that I can correctly assert (at least according to some idiolects of English), "I remember that I ate pizza for breakfast last Sunday," even if I believe that I didn't. Admittedly, it might be more felicitous in such cases to say something like, "I *seem to* remember that I ate pizza for breakfast last Sunday," but the rephrasing is not mandatory for at least some speakers. Thus, factivity is not an obligatory feature of our commonsense concept, and I would argue, that it ought not to be a desideratum for the scientific category of *episodic memory*.[28]

---

[28] Craver (2020) argues that our concept of memory ought to be bifurcated into epistemic and empirical concepts. But I submit that the need to bifurcate is greatly diminished if factivity is not regarded as a feature of memory.

CHAPTER 6

Language-Thought Processes

The limits of my language mean the limits of my world.
— Wittgenstein, *Tractatus Logico-Philosophicus*

When a language dies, so much more than words are lost. Language is the dwelling place of ideas that do not exist anywhere else. It is a prism through which to see the world.
— Robin Wall Kimmerer, *Braiding Sweetgrass: Indigenous Wisdom, Scientific Knowledge, and the Teachings of Plants*

6.1 Introduction

This chapter will look at a type of cognitive process that has been studied by linguists and other cognitive scientists at least since the early twentieth century. Though this type of cognitive process is not explicitly regarded as a cognitive kind, it can be considered to be a candidate for being such a kind, and can be evaluated as such. The type of process I have in mind is one that involves a particularly strong influence of language on thought, or a deep and far-reaching relationship between language and thought. Whether or not there are such cognitive processes is itself a matter of debate, and their very existence has been hotly contested. The issue tends to be framed in terms of subscription to a hypothesis, the "Sapir-Whorf hypothesis," rather than the existence of a particular kind, but in the context of this book, I will consider it to be a question regarding the existence of a particular *kind* of cognitive process, as I will try to explain in the course of this chapter.

The "Sapir-Whorf hypothesis," or sometimes just plain "Whorfianism," is one of the most widely discussed general claims about language. It has attracted some limited attention among philosophers, mainly because of its connection to broader philosophical concerns having to do with conceptual change and conceptual incommensurability (Davidson 1974; Carruthers 2002). Meanwhile, although psychologists and linguists have

6.1 Introduction

investigated the thesis extensively in the past few decades, there does not seem to be a consensus on its actual content or how to formulate it. In this chapter, my aim is to formulate the thesis with greater precision than it appears to have been stated thus far. This will enable us to determine whether it demarcates a real kind of phenomenon or phenomena, and whether the phenomena in question are categorically distinct from others that involve the interaction of language and thought. That, in turn, should help us assess its status as a real cognitive kind. So as not to get embroiled in an exegetical dispute about the actual views of its most famous proponents, Edward Sapir and Benjamin Lee Whorf, I propose to employ a more neutral handle for the thesis that has often been associated with their names, the "Language-Thought (LT) hypothesis." The corresponding candidate for a cognitive kind can be labelled a "LT process" and the psychological effects that are thought to be accounted for by this process can be called "LT effects."

In its simplest form, the LT hypothesis is often stated as the claim that language influences thought or cognition. The thesis is often given in a weak and strong version; sometimes the weak version is labeled "linguistic relativity," while the strong version is termed "linguistic determinism." The weak version is often taken to be the bare claim that language influences thought (to some degree),[1] while the strong version is widely held to be the view that language determines thought (cf. Kay and Kempton 1984; Clark 1996). There is a problem with both versions of this formulation of the LT hypothesis. The former claim is innocuous or indeed trivial: How could there not be some influence of language on thought? (Compare: the view that perception influences thought, or the thesis that emotion influences cognition.) By contrast, the latter claim seems ridiculously inflated: How could language alone determine a person's or a community's entire body of thought or cognitive processes? But if neither of these statements of the position are tenable, is there an intermediate view that is both interesting and plausible, that is neither trivially true nor patently false?[2] The problem has been made more acute by a challenge laid down by two cognitive scientists who have dramatically highlighted the threat of trivialization when it comes to the LT hypothesis. Bloom and Keil (2001) argue that the issue

---

[1] "Linguistic relativity" is a misleading label for this position, since it suggests merely that languages differ in the way that they describe or categorize features of reality, without making the additional claim that these differences have an impact on thought.

[2] The lack of consensus on the precise content of the thesis is confirmed by the fact that researchers sometimes reach opposite conclusions regarding the support for the thesis based on much of the same evidence (compare e.g. Bloom and Keil (2001) with Reines and Prinz (2009)).

in question cannot be a matter of whether language can have an influence on thought, since: "Nobody doubts that language can inform, convince, persuade, soothe, dismay, encourage, and so on" (Bloom and Keil 2001, 354). Otherwise, they ask rhetorically, "why would you be reading this?" It is an obvious truth that language influences thought every time one cognitive agent uses language to communicate thoughts to another, and this fact threatens to trivialize the LT hypothesis unless we can distinguish this ubiquitous feature of human cognition from the phenomena that are supposed to support the hypothesis.

If one does not take one's cue from the actual views of Sapir and Whorf, how should one proceed to pin down the content of the LT hypothesis with sufficient rigor as to avoid the threat of trivialization and determine whether it has correctly managed to identify a real cognitive kind? I propose to be guided by two sources of evidence, recent empirical work claiming to test the hypothesis, which contains an implicit understanding of its content, and current attempts to state the thesis in a rigorous way. By building on ideas and insights gleaned from these two sources in turn, I aim to formulate it more explicitly and precisely than it has usually been stated. It might seem wrong-headed to survey the empirical evidence before one has looked at attempts to formulate the hypothesis. How can we tell whether the evidence is indeed evidence for the LT hypothesis if we do not yet have a formulation of that hypothesis? The answer is that the empirical cases that I will examine are widely cited as archetypal pieces of evidence for the hypothesis and contain an implicit, if not explicit, understanding of that hypothesis. If it turns out, upon reflection, that some of the evidence cited does not conform to the formulations that we will later encounter, then that evidence can be reassessed, or else the formulations themselves can be revised. The point of the exercise is not merely descriptive, since I will be arguing that some ways of stating the thesis and some implicit understandings of the processes involved are more precise, cogent, and empirically corroborated than others.

The chapter will proceed as follows. In Section 6.2, I will survey some of the empirical evidence that is widely cited to support the LT hypothesis. Then, in Section 6.3, I will examine a number of attempts to formulate the hypothesis, comparing them, pointing out various shortcomings, and highlighting the most promising aspects of each. In Section 6.4, guided both by the empirical results in Section 6.2 and the formulations surveyed in Section 6.3, I will attempt to frame the LT hypothesis in some detail and with sufficient rigor, in such a way that it evades the charge of trivialization and can be used to describe a candidate for a *kind of cognitive process*. Though it will emerge that there is significant support for the LT hypothesis in recent empirical work, I will argue that the processes demarcated by the LT hypothesis may not be entirely distinct from other types of cognitive process. I will also make the case that they may not constitute a homogeneous collection of cognitive processes in their own right. The upshot will be that Language-Thought Processes do not constitute a cognitive kind, but rather that they are of two distinct types of process, each of which may be subsumed within a broader cognitive kind.

6.2 Empirical Evidence for the LT Hypothesis

Writing a few decades ago, Kay and Kempton (1984, 67) asserted that "the bulk of the [empirical] research" designed to test the LT hypothesis "has concerned the domain of color." Fortunately, this assessment has been made obsolete in the interim, since the past few decades have witnessed a flurry of empirical activity that has examined the LT hypothesis in a number of different domains. Consider a few paradigmatic experimental results that are meant to support the LT hypothesis, drawn from different areas of cognition:

a) ***Color:*** Russian speakers, who use different terms for light blue (*goluboy*) and dark blue (*siniy*), can more easily discriminate two shades taken from the two distinct categories than two shades taken from the same linguistic category; English speakers do not show this effect. Thus, linguistic categories affect performance on a nonverbal perceptual task. The effect for Russian speakers can be disrupted by a verbal interference task, apparently confirming the influence of language on thought (Winawer, Witthoft, Frank, et al. 2007).[3]

b) ***Spatial coordinates:*** Speakers of languages that primarily employ absolute spatial coordinates (e.g. Guugu Yimithirr, Tzeltal) carry out nonlinguistic spatial tasks differently than speakers of languages that primarily employ relative spatial coordinates (e.g. English, Dutch). Specifically, the former tend to duplicate visual scenes and remember them in a way that accords with cardinal directions, while the latter do so in accord with egocentric directions (Levinson 2003; Majid, Bowerman, Kita, et al. 2004).

---

[3] Since color is perhaps the most active research focus in this broad area of inquiry, there is obviously a great deal more to be said about the influence of color language on color perception. See Cibelli, Xu, Austerweil, et al. (2016) for a model of color cognition that combines a universal color space with language-specific partitions of that space.

c) **Gender:** Spanish and French speakers show effects of grammatical gender on classification when asked to assign either a woman's voice or man's voice to an object in a picture (supposedly for an animated movie). These effects emerge reliably only around 7 years of age, indicating that there is a "time lag between language acquisition and the infiltration of language into the cognitive system ..." of the order of several years (Sera, Elieff, Forbes, et al. 2002; cf. Boroditsky, Schmidt, & Phillips 2003).

d) **Categorization:** English-speaking and Japanese-speaking children and adults generalize object instances and substance instances differently, suggesting an influence of language on categorization, having to do with differences in the use of count-mass terms and classifiers in English and Japanese. The differences are more pronounced among adults than among children (Imai & Gentner 1997).

e) **Motion:** Direction of motion is more often conveyed by the verb itself in Spanish and by a preposition in English, whereas manner of motion is usually encoded in the verb itself in English and in other supplemental expressions (e.g. adverbs) in Spanish. Correspondingly, children who speak English describe an illustration from a storybook differently from their Spanish-speaking counterparts, and the effect is more pronounced at age nine than at age five. It has been suggested not only that native English and Spanish speakers describe a visual scene differently due to linguistic differences between the two languages, but that the process of "thinking for speaking" eventually leads children to notice different things about a visual scene (Slobin 1996; cf. Slobin 2003).

Although all these results have been widely cited, none of them are without their critics and detractors. But the criticisms tend to claim that the evidence does not adequately support the conclusion, not that the conclusion is incoherent or not well formulated. For example, the study cited in example (e), above, has been criticized on the grounds that it does not show that children notice different things in a picture but that they simply describe what they notice differently (Gennari, Sloman, Malt, et al. 2002, 55). However, where these results have been accepted, they have been taken to demonstrate LT effects. Hence, it seems warranted to regard them at least as a tentative guide to what genuine LT effects would amount to, and hence what a LT process would be.[4]

---

[4] There is another class of effects and body of research that pertain to a generalized version of the hypothesis that language influences thought. This version states that having language endows humans with

6.3 Formulations of the LT Hypothesis

Now that we have sampled some representative empirical results that have been judged to support the LT hypothesis, we should have a better idea of what we are looking for. The common denominator among these results would seem to be that they all show evidence that the native language[5] of a speaker has some influence on other, nonlinguistic aspects of the speaker's cognitive capacities (e.g. perception, memory, spatial cognition, categorization). The influence is, presumably, causal, and the effects are supposed to obtain even when these aspects of a speaker's cognitive capacities are tested on tasks that do not require explicit linguistic representation. To be sure, in some of the experiments mentioned in the previous section, language is involved in some way in performing the task, but the aspects of language that are under examination are assumed not to be explicitly recruited by the task demands.

There are few attempts in the literature to formulate the LT hypothesis in any detail; many research articles on the topic content themselves with a paraphrase to the effect that language influences thought or nonlinguistic cognition in some way. Indeed, some formulations of the LT hypothesis resort to figures of speech, employing the metaphors of language "shaping" or "molding" thought, without attempting to describe the type of process in any detail. The four formulations to be discussed here constitute exceptions in that they attempt to articulate the LT hypothesis in more detail and in a nontrivial fashion.

---

certain cognitive capacities that they would have otherwise lacked. This general linguistic capacity (as opposed to specific natural languages) is supposed to be at least partly responsible for some of the cognitive abilities of human beings that set them apart from other animals (see e.g. Dennett 1997; Clark 1998; Gentner 2003; Spelke 2003). This thesis seems weaker than the LT hypothesis, in the sense that one could hold it while not accepting LT, though the converse may not be true. If one holds that having language, as opposed to having no language at all, has an influence on thought, it does not follow that mastering, for example, Arabic as opposed to English has an influence on one's thought. By contrast, if it is true that being a native or competent speaker of some specific language has such an influence, then it would follow that having language generally influences thought. Since it is a weaker thesis, I will not discuss the general thesis further here, though formulating the specific LT hypothesis may also help to articulate the precise content of the more general thesis. In Chapter 2, I discussed the influence of language on concept possession and acquisition, arguing that linguistic symbols are instrumental in determining conceptual identity, and this point will be taken up again in Section 6.4.

[5] Why does it have to be one's native language? Presumably because effects associated with one's native language are more far-reaching than those associated with other languages that one might have mastered. Hence, if there is an effect at all, it ought to be more noticeable in the case of a thinker's native language. But see, for example, Dolscheid, Shayan, Majid, et al. (2013) for LT effects resulting from training native Dutch speakers on a linguistic metaphor in Farsi. But they posit that "participants received a concentrated 'dose' of the relevant linguistic metaphor, probably equivalent to weeks or months of normal language use" (Dolscheid, Shayan, Majid, et al. 2013, 619). Thus, these effects may pertain more properly to proficient or competent speakers rather than native speakers.

6.3.1 Bloom and Keil

In an attempt to respond to their own challenge regarding the trivialization of the LT hypothesis, Bloom and Keil (2001, 354) state that the real question is not whether language can have an influence on thought, since this occurs every time we use language to communicate with one another. Rather, the issue is "whether language shapes thought in some way other than through the semantic information that it conveys," that is, "whether the structure of language – syntactic, morphological, lexical, phonological, etc. – has an effect on thought." But this response does not seem quite right for two reasons. First, LT effects do not always involve structural differences among languages, even if we understand linguistic structure more broadly than just in terms of syntax. In the case, for example, of the experiments concerning Russian speakers' perceptual discrimination of shades of blue, the influence does not have to do with the structure of the Russian language, merely with having two words to denote different shades of blue (example (a) in Section 6.2). At the very least, it is debatable whether we should consider that Russian's having an additional lexical item relative to English is a difference in lexical structure. The second problem with Bloom and Keil's proposal is that these linguistic differences often do involve semantic information. A Russian speaker who has an additional lexical item for a shade of blue can be considered to have additional *semantic* information, for example, she knows that *siniy* refers to *that* shade. Similar remarks apply to languages with absolute as opposed to relative coordinates (example (b) in Section 6.2), which supply language users with additional conceptual resources, without affecting the very structure of the language.[6] But there may be a way of modifying Bloom and Keil's proposal to make it fit the phenomena. In all the cases sampled in the previous section, the speaker is equipped with one or more additional lexical or morphological items to stand in for certain concepts or conceptual constituents. Rather than say that the structure of language influences thought in these cases, it may be more accurate to say that certain aspects of language (primarily lexical or morphological aspects) influence certain habits of thought or cognitive tendencies. In fact, the issue does not appear to be about whether the *structure of language* influences thought, but rather whether language influences the *structure of thought*, such as categorization abilities or perceptual discrimination. But

---

[6] Even examples (c), (d), and (e) in Section 6.2 are not purely syntactic in nature, but combine syntax and semantics. It is true that in some of Sapir and Whorf's earlier work, syntactic differences were emphasized over semantic ones, but the bulk of the empirical work since then suggests that both kinds of LT effects are attested, and indeed are hard to disentangle.

6.3 Formulations of the LT Hypothesis

this constitutes a rather vague response to Bloom and Keil's challenge, so I will try to make these locutions more precise in what follows.

6.3.2 Carruthers

Another attempt to delineate the phenomena of interest comes from Carruthers (2012), who observes that traditional "Whorfianism" held that language has a "*structuring* effect on cognition," thereby seeming to confirm the variation on Bloom and Keil's formulation proposed in Section 6.3.1. But he goes on to explicate this idea by saying that it holds that "the absence of language makes certain sorts of thoughts, or certain sorts of cognitive process, completely unavailable to people" (Carruthers 2012, 385). This claim, which Carruthers rejects, accords with the strong version of the LT hypothesis mentioned earlier, which states that language determines thought in the sense that it prohibits (or mandates) the thoughts that are available to speakers. After rejecting this discredited view, he goes on to argue that a more plausible view has recently been gaining ground, namely that "natural language can make certain sorts of thought and cognitive process more *likely*, and more *accessible* to people" (Carruthers 2012, 385; original emphasis). The idea that language makes certain thoughts or cognitive processes more likely and accessible appears to accord with the experimental results vetted in Section 6.2. Indeed, Carruthers to the contrary notwithstanding, this would seem to be one way of elaborating on the vague notion that language "structures" cognition. One obvious way to test for whether cognition has been restructured is to gauge whether certain cognitive processes have been rendered more likely to occur and more accessible to language users. If our cognitive abilities have indeed been restructured, then this should show up in the cognitive processes that we engage in, especially in making certain responses more or less likely, or in making certain thoughts more or less accessible. Hence, we might do well to retain the idea of language restructuring thought, at least provisionally, in attempting to formulate the LT hypothesis, when understood in terms of making some thoughts more likely or accessible than others. This is also consistent with the claim that the influence of language on thought does not involve either mandating or prohibiting certain thoughts or habits of thought outright. In other words, it agrees with the widespread consensus that to the extent that the LT hypothesis is defensible, it is so in its weak rather than its strong form. In contrast with strong LT, weak LT talks about facilitating (or inhibiting) rather than mandating (or prohibiting) cognitive processes, in the sense of making them more or less likely or more or less accessible.

6.3.3 Hunt and Agnoli

In a somewhat earlier discussion, which remains relevant, Hunt and Agnoli (1991) put forward some concrete suggestions for thinking about the ways in which language might restructure thought. "The weaker form of the [LT] hypothesis states that language differentially favors some thought processes over others, to the point that a thought that is easily expressed in one language might virtually never be developed by speakers of another language" (Hunt & Agnoli 1991, 378). This agrees with the idea encountered above (in Section 6.3.2) that some thoughts may be rendered more probable and accessible for speakers of some languages as opposed to others. Hunt and Agnoli expand on this by saying that some thoughts will be more "natural" and "come easily" for the speakers of some languages relative to others, though they admit that these notions are not capable of a scientific construal. However, they go on to make some interesting suggestions as to how one might render the contrast more amenable to empirical investigation. Using the example of southern Californian surfers who have recruited words like "hollow" and "flat" to describe specific types of ocean waves, Hunt and Agnoli (1991, 378) claim that a language user who lexicalizes a concept has "traded expensive space in short-term memory for cheaper space in long-term memory." Presumably, even though non-surfers may be able to recognize the relevant differences among waves and recall them on a later occasion, surfers may be more efficient at doing so, partly because they have convenient labels that can be recruited, whether overtly or covertly, to do so. Lexicalizing the concept in a language may make it more probable that the concept is accessed by a thinker, or make it accessible more efficiently. Moreover, it may make it possible to access it while performing other, nonlinguistic cognitive tasks that make demands on short-term memory. This, then, may be one process whereby cognition may be restructured by language, or some thoughts rendered more probable and accessible as a result of the introduction of linguistic expressions for them.

But though this is an illuminating discussion, it does not yet answer the Bloom-Keil challenge, since the question still remains: How do these kinds of facilitatory effects differ from the ways in which language routinely promotes or encourages thoughts in ordinary human communication? It may be useful to compare two cases of human communication to better understand the difference. Consider first a case in which I ask a first-grader to turn the lights off when she leaves the bathroom after brushing her teeth. She assents, and minutes later, she complies with my request. Here, a linguistic utterance has influenced the thoughts of my interlocutor, leading to her subsequent action. Now imagine a case in which I explain to the same first-grader that the planet's climate is changing as a result of human activity, perhaps introducing her to such concepts as *climate* and *global warming*. Then that may encourage her to think certain thoughts, to the effect that she should take steps to reduce her own energy use and that of others. It may make it more likely for her to think of switching off the lights when she leaves a room, not just on a specific occasion but on many future occasions. It may even entail a significant restructuring of her thoughts about her relationship to her environment, her consumption habits, diet, lifestyle, and so on, prioritizing thoughts about energy conservation as she makes daily choices about transportation, eating, and entertainment. In both cases, I have used language to influence the thoughts (and actions) of the child. But it is clear that in the second case, the influences are more far-reaching and enduring. There may be no strict dividing line between the first type of case and the second, and there may be intermediate cases (for example, a case in which I explain to the child that it is a waste of energy to keep lights on in an empty room without going into the ramifications for the planet's climate). But it seems clear that there is a significant difference between the first type of case and the second, and that part of the difference has to do with lexicalizing concepts like CLIMATE and GLOBAL WARMING. Once these concepts are lexicalized, they are stored in long-term memory, as Hunt and Agnoli observe, and may have a more significant impact on the cognitive life of the agent in the long run, making some thoughts generally more likely and more accessible.

It may be objected that in the case of explaining climate change, the child is not learning a natural language but is instead learning a scientific theory, and so the effect on cognition results from acquiring a specific body of knowledge, not simply from being proficient in a natural language. Hence, we might stipulate that LT effects are those that affect a language user merely in virtue of being a competent speaker of that language, not in virtue of learning a new theory or a new set of concepts in that language. Part of what distinguishes LT effects from standard communicative acts may be said to be that they involve mere competence in a cognitive agent's natural language, rather than familiarity with a specialized domain of knowledge. In genuine LT effects, it appears that it is important that the effect in question result merely from the fact that the language speaker is a competent speaker of his or her native language. (However, this assumption will be reexamined in Section 6.4.)

6.3.4   Wolff and Holmes

Finally, a more recent formulation of the LT hypothesis comes from Wolff and Holmes (2011, 261), who use a metaphor to convey the gist of the thesis: "There is evidence … that while language may not close doors, it may fling others wide open. For example, language makes certain distinctions difficult to avoid when it meddles in the process of color discrimination or renders one way of construing space more natural than another." This suggests that the LT hypothesis construes language as a promoter (or inhibitor) rather than a dictator (or prohibitor), making some thoughts more or less likely (rather than mandatory or unattainable), as we have already concluded. But (again) this is not sufficient to answer the challenge conjured up by Bloom and Keil. Wolff and Holmes (2011) go on to provide a useful taxonomy of the ways in which language can "open doors." The main categories in their taxonomy are as follows: (i) thinking before language, (ii) thinking with language, and (iii) thinking after language, and they provide examples of each of these categories. In the first category, they place the phenomenon of "thinking for speaking," according to which, for example, English- and Spanish-speaking children notice different things about a picture because of the need to put it into words in their respective languages (Slobin (1996); example (e) in Section 6.2). In the second category are such phenomena as the color discrimination tasks with Russian and English speakers, in which Russian speakers are better at discriminating color shades that are labeled differently in Russian (Winawer, Witthoft, Frank, et al. 2007; example (a) in Section 6.2). Meanwhile, the third category includes cases involving differences between speakers of languages with absolute coordinates and speakers of languages with relative coordinates, specifically differences in how they replicate a visual scene (Majid, Bowerman, Kita, et al. 2004; example (b) in Section 6.2). This classification scheme is based mainly on the temporal sequence in which language and thought interact, whether the influence of language on thought occurs before, during, or after the cognitive process. But that cannot be quite right, since it is incoherent to say that language has an influence on the cognitive process *after* that process has occurred. Rather, if one lumps together (i) and (ii), and contrasts them with (iii), the distinction seems to be between two kinds of influence, one that is more time-sensitive, in which language is covertly implicated in a particular thought process as it occurs or shortly before it occurs, whereas the other has to do with changes in cognition that have been effected over time as a result of habitual language use and persist even in the absence of language use. In the first two broad categories, the interaction between thought and language is simultaneous or nearly so. Even when performing some nonverbal tasks, there is evidence that thinkers use language as a tool or crutch and that it induces them to think certain thoughts or engage in certain cognitive processes. The paradigmatic examples of these cases may be those that involve lexicalizing a concept, allowing us to store it in long-term memory and access it more reliably in carrying out certain cognitive tasks, such as perceptual discrimination tasks. In the third category are phenomena that implicate ingrained habits of thought that are the result of linguistic influences that are more long-term than the other two. These cases would seem to be different in that they typically involve associations among concepts, which are made as a result of habitually using a language and of making certain distinctions required by some natural languages (though not others). Using a language that makes gender distinctions may encourage speakers of that language to make associations between gendered nouns for inanimate objects and stereotypical properties associated with males and females in that society (as in example (c) in Section 6.2). Therefore, there would appear to be at least two kinds of cognitive process involved in typical LT effects. The first, which corresponds roughly to (i) and (ii) in the above taxonomy, involves long-term memory storage of certain distinctions that are lexicalized in one's native language. The second, corresponding to (iii), involves making habitual associations between aspects of reality that are associated by one's native language. As Wolff and Holmes observe, some LT effects are typically erased or diminished by verbal interference but others are not. The reason is that in the type-(i) and type-(ii) cases, language is thought to be covertly recruited in performing the task, whereas in the type-(iii) cases, language need not be recruited simultaneously, since it is posited to have had the effect on cognition after a long period of use, and the effect is not cancelled by temporarily disrupting our linguistic cognitive processes through verbal interference tasks. These latter cases may include those experimental results surveyed in Section 6.2 in which the effect appeared or became more pronounced only several years after learning language, or even in adulthood (examples (c), (d), and (e)).

    A great deal more work would have to be done to understand the processes that underlie these effects, but at a first pass, there are likely to be two distinct kinds of process responsible for these two types of LT effects. In the first kind of process, long-term storage is involved, whereby linguistic labels enable us to store items in long-term memory rather than having to keep them in short-term memory, thereby saving precious cognitive resources. This would explain why some items can be recalled more efficiently and reliably. It may also explain why linguistic interference would disrupt this task, since one's language faculty may need to be recruited to facilitate performing the recall task, which is mediated by the use of a linguistic label. As for the second kind of process, it is likely to be of the associative sort, which appears elsewhere in cognition. In this case, frequent association between concepts, due to language use, leads to a strong connection between these concepts, even when the concepts are not recruited by a task involving language use. This distinction will be explored further in the next two sections.

6.4 Proposal and Discussion

Now that we have surveyed several prominent recent attempts to pin down the LT hypothesis, it is possible to derive the most promising elements from these attempts in order to try to emerge with a more precise formulation. The LT hypothesis can be said to consist of the conjunction of the following four claims concerning particular natural languages:

1) Language *facilitates or inhibits* (rather than dictates or prohibits) a speaker's thoughts or cognitive processes;
2) Language effects *structural cognitive modifications*, making certain thoughts more (or less) accessible, in cognitive tasks involving perceptual discrimination, memory, and categorization, among other cognitive processes;
3) Language is implicated in the effect merely as a result of the agent being a *competent speaker* of that language, not as a result of acquiring additional information, such as expert knowledge;
4) Language (i) has a *near simultaneous* and covert causal influence on cognition as a result of lexicalizing one or more concepts, or (ii) produces a *nonsimultaneous and long-term* causal change in cognition as a result of habitual associations.

Even if they do not constitute necessary and sufficient conditions, these can be considered the main tenets of the LT hypothesis, and the paradigmatic experimental results sampled in Section 6.2 seem to bear them out. But do they enable us to answer the Bloom-Keil challenge, and do they delineate a homogeneous and distinct class of cognitive effects? When it comes to the former question, some of these clauses, particularly (3), rule out the kind of communicative processes by means of which language routinely affects thought as in standard communicative acts. But it may be objected that we have, in effect, *defined* the phenomenon in such a way

6.4 Proposal and Discussion

that the Bloom-Keil challenge does not arise. Perhaps, but the initial task was to delineate this class of effects, not to justify why these effects ought to be distinguished from other cognitive phenomena involving language. As I will try to argue presently, there is unlikely to be a principled difference between those cognitive effects that result from being a competent speaker and those that ensue from acquiring expert knowledge. This formulation of the LT hypothesis may accord with the paradigm cases and may indeed capture those phenomena that have commonly been considered LT effects, but that does not mean that LT effects constitute either a homogeneous or a distinct category of cognitive phenomena. Now that we have a delineation of the phenomena, we can go on to consider whether there are principled reasons for focusing on this class of effects. Do they constitute a homogeneous and distinct category of effects, in other words, are they the effects of a *kind* of cognitive process? In the rest of this section, I will try to argue that not only are LT effects as a whole not likely to be distinct from a broader set of cognitive effects, they are also likely to issue from two different kinds of cognitive process.

It will be convenient to proceed by considering each of the claims (1)–(4), above, starting with (1), the idea that LT effects are facilitatory or inhibitory, rather than mandatory or prohibitory. This aspect of the LT hypothesis reflects the consensus that the version of the hypothesis that has strongest empirical support is the weaker rather than the stronger version, sometimes known as "linguistic relativity" (as opposed to "linguistic determinism"). The existing evidence suggests that whatever cognitive effects language may have, they neither mandate nor prohibit certain thoughts or cognitive operations. Some philosophers and cognitive scientists do claim that some languages are incommensurable with others, but they would still allow that thinkers can make the transition from one language to the other or learn the second language (Carey 2009). Presumably, these language learners can acquire concepts that are not lexicalized in their native language, even though they may have to overcome certain cognitive hurdles to do so. Hence, (1) is a fairly well-established component of the LT hypothesis and seems to be supported by the effects we have surveyed.

As for (2), the claim that language effects certain structural modifications in one's cognitive system, this has been explicated in terms of making certain thoughts more likely or accessible (though that may be understood as a way of testing or operationalizing the claim). One process for achieving a restructuring is the one proposed by Hunt and Agnoli (1991), who contend that lexicalizing a concept may involve (among other things) storing it in long-term rather than short-term memory.[7] This would facilitate access to it and ensure that it can be used in a cognitive task that places demands on short-term memory, as in tasks involving remembering the spatial configuration of objects (example (b) in Section 6.2). But there may also be other cognitive processes that yield LT effects, such as habitual associations built up between concepts as a result of the fact that the corresponding words or morphemes are associated in natural language. For instance, French speakers may associate the word for an inanimate object that takes the masculine gender with stereotypical masculine traits and hence be more prone to associate that object with a male voice rather than a female voice (example (c) in Section 6.2). Speakers of languages in which the word for the same object takes the feminine gender or is gender-neutral may not exhibit the same tendency because they have not formed the habitual association between the object in question and the stereotypical gender traits. The processes that subserve these associative effects seem different from those that would be involved in storing a lexical item in long-term memory. They need not be disrupted by verbal interference and may take longer to become entrenched. Despite the different processes involved, both types of LT effects can be thought of as restructuring cognition in the sense of making certain cognitive responses more accessible or more likely.

When it comes to clause (3), this may be the most controversial aspect of the above characterization of the LT hypothesis. This clause was used to distinguish standard LT effects from instances of concept acquisition leading to cognitive restructuring, as in the case of a child whose habits of thought change as a result of learning about climate change, acquiring such concepts as CLIMATE and GLOBAL WARMING. We might distinguish these phenomena from LT effects by saying that they do not involve the acquisition of a natural language, but rather the acquisition of a novel set of concepts within a language. But the problem with this suggestion is that the line between languages (or language fragments) and conceptual repertoires may not be a principled one. Indeed, some of the examples that have been used to illustrate the phenomenon of cognitive restructuring as a result of LT effects involve the acquisition of concepts by speakers within the same natural language, as in the example of California surfers lexicalizing certain wave formations (mentioned in Section 6.3.3), rather than a contrast between speakers of different natural languages. It is well

---

[7] Presumably, it also somehow involves establishing a link between long-term semantic memory and a particular lexical item or morpheme, since lexicalization is not a mere matter of storage in semantic memory.

6.4 Proposal and Discussion

established that within the same natural language, experts tend to deploy concepts and categories that novices do not, for example, for describing different tastes of wine (Solomon 1990), or for classifying different types of physics problem (Chi, Feltovitch, & Glaser 1981), among many other features of reality. LT effects do not seem different in principle from expert-novice effects, for these also influence perception, memory, and other nonlinguistic cognitive functions, as is now well established, and as some proponents of the LT hypothesis have themselves observed (cf. Majid 2002). Moreover, they would appear to involve the same process of lexicalization leading to storage in long-term memory. The difference between the deployment of additional concepts by experts (compared to novices) within a single natural language and the deployment of distinct concepts by speakers of different natural languages is not a principled one from the psychological or linguistic point of view. Indeed, if the linguistic differences between experts and novices are pronounced enough, they can be considered to speak different dialects, and it is a platitude that the difference between languages and dialects is hardly principled either. This seems to leave us with no clear way of distinguishing at least some LT effects from some expert-novice effects in which concepts are lexicalized in a specialized vocabulary. What seems more relevant is the extent to which the acquisition of information leads to cognitive restructuring (as in clause (2) above). This kind of restructuring is far more likely to result from acquiring new concepts or a new theory (e.g. about global climate change) rather than the simple exchange of information in an ordinary communicative act. But the restructuring can just as well be the result of acquiring and lexicalizing novel concepts as it can be the result of mere competence in a natural language.

A different sort of problem with this characterization of LT effects has to do with clause (4), which would seem to pick out two distinct classes of phenomena, those in which language is simultaneously and covertly recruited during a cognitive process and those in which language has a long-term effect on cognition due to habitual association. Now there are some researchers who tend to regard the former as not being genuine LT effects, and some experimental results are summarily dismissed or explained away by saying that they involved "inner speech" or some unobserved recruitment of the linguistic capacity during the completion of a cognitive task (see e.g. Bloom & Keil 2001, 356; cf. Casasanto 2008, 70; Casasanto 2016, 159–160). Other researchers hold that explicit recruitment of language during the completion of a nonlinguistic task is at least as significant as habitual associations that are reinforced by an extended period of language use (see e.g. Lupyan, Abdel Rahman, Boroditsky, et al. 2020). But there is no need to discuss which of these two types of effect is more important, since the crucial point for our purposes is that there appear to be two types of cognitive process in play, corresponding to two different putative cognitive kinds. In the next section, I will try to further articulate the difference between these two cognitive processes, based on current empirical findings and taxonomic practices in the cognitive sciences.

6.5 Two Kinds of LT Process

Over the past decade, empirical work on LT effects in cognitive science seems to be converging on the conclusion that there are two distinct types of cognitive process involved in producing such effects. A number of researchers have tried to articulate the differences between these two kinds of cognitive process, as can be seen from the following remarks by contributors to this research program:

> … language affects performance differently in different cases. In the case of color discrimination [e.g. Winawer, Witthoft, Frank, et al. 2007], language appears to be having an acute effect; it is involved online, meddling in perceptual decisions as they are made. In the case of grammatical gender [e.g. Boroditsky, Schmidt, & Phillips 2003], language appears to be more of a chronic affliction; it has had a long-term effect on the underlying conceptual representations such that even if language is disabled, its effects in the conceptual system remain. (Boroditsky 2012, 625)

> One possibility is that these [cross-linguistic perceptual] differences stem from long-term perceptual learning caused by years of distinguishing colors in one language … An alternative is that the cross-linguistic perceptual differences arise from online top-down influences of language … (Perry & Lupyan 2013, 122)

> The present results support the proposal that language can also influence people's low-level perceptuomotor abilities … and that cross-linguistic differences in mental representation can be observed even when people are not using language on-line, overtly or covertly. (Dolscheid, Shayan, Majid, et al. 2013, 619)

> We begin by describing two types of linguistic effects on perception: off-line effects in which long-term experience with a specific language affects how people subsequently experience certain perceptual inputs, and on-line effects in which some aspect of language, such as an object's name, interacts with in-the-moment visual processing. (Lupyan, Abdel Rahman, Boroditsky, et al. 2020, 931)

These observations confirm that current empirical evidence points to the conclusion that all language-thought effects should not be lumped together in the same category, but rather that they divide into (at least) two distinct taxonomic categories (an argument also made by Wolff & Holmes (2011), as seen in Section 6.3.4).

In discussing the various clauses of the formulation of the LT hypothesis in Section 6.4, we have been led to acknowledge that it neither delineates a class that is distinct from other cognitive phenomena (i.e. ones not usually classified as LT effects) nor does it pick out a homogeneous class of cognitive phenomena. It does not pick out a distinct class because the cognitive effects of conceptual disparity across natural languages are not in principle different from those of conceptual innovation within the same natural language. A Californian surfer equipped with additional terms to describe different types of ocean waves is not in principle different from a Russian speaker who has two terms to describe different shades of blue, relative to a speaker of (standard) English. Still, Bloom and Keil's trivialization problem is not vindicated, since the cognitive effects associated with LT phenomena are not akin to simple communicative acts using language, but are perhaps more closely aligned with those of conceptual change or conceptual enrichment, as in the acquisition of new scientific concepts or a novel set of expert categories. We cannot summarily dismiss these phenomena as non-LT effects by adding a clause to the effect that LT effects involve mere competence in a natural language, since there is no clear distinction between linguistic competence in a natural language and mastery of specialized vocabulary in that language. But in addition to the fact that LT effects may not be clearly demarcated from other cognitive phenomena, these effects may not be coherent or homogeneous, since those effects that involve the simultaneous recruitment of language ("online") are likely to result from different cognitive processes than those that do not ("offline"), though neither of these processes is currently understood well enough to enable us to say for sure. The terms "online" and "offline" are convenient labels that have been used in some recent work to denote the difference between these two kinds of causal process (e.g. Casasanto 2016).[8] But though they are suggestive metaphorical terms, they should not hide the fact that the precise causal pathways in question are not yet well understood. Still, it may be worth summarizing a few of the main features of each type of process, as currently conceived.

---

[8] See also Boroditsky (2012), Lupyan (2012), Perry and Lupyan (2013), Dolscheid, Shayan, Majid, et al. (2013), and Lupyan, Abdel Rahman, Boroditsky, et al. (2020), most of which were quoted above. But note that Slobin (2003) seems to use the term "online" for processes that others would label "offline."

The first type of cognitive process entails lexicalizing a concept and storing a lexical item in long term memory. This item is then thought to serve as a tag or label that is covertly deployed in the course of performing some nonlinguistic task, such as a perceptual or memorial task, facilitating the performance of that task. Lupyan (2012) has proposed the "label-feedback hypothesis" to account for this type of language-thought effect, though he admits that it is just a sketch. As he explains it, visual representation of the color blue, say, can spontaneously activate the word "blue," and this in turn modulates the visual representation. This means that linguistic labels can be covertly deployed in the course of performing some nonlinguistic task, often providing a way of carrying it out more accurately or rapidly, or at any rate differently from the way in which it would be carried out without dependence on the linguistic label. Some researchers think that this may involve "subvocal linguistic encoding" during the performance of the task (Bohnemeyer 2020), while others posit that the "phonological loop," widely considered to be a constituent of short-term or working memory, has a role to play in executing such tasks (Athanasopoulos & Casaponsa 2020). This is why such effects are thought to be extinguished under conditions of verbal interference.

The second type of cognitive process typically operates over a longer time period, since it involves the forging of habitual associations between two or more concepts as a result of prolonged language use. Different languages represent concepts differently. In the simplest case, some languages may lexicalize certain concepts, while others may not. This means that over a long period of language use, the features associated with those concepts will be made more salient for users of some languages rather than others, and they will tend to notice them more frequently, be more efficient at categorizing instances falling under them, and otherwise perform differently on certain cognitive tasks. Moreover, these tasks need not involve language use, whether overt or covert, as long as they result in significant effects on overall cognition. Similarly, some languages make associations between concepts that other languages do not. For example, while Dutch describes acoustic pitches as high or low, Farsi describes them as thin or thick, thus forging associations between these acoustic and spatial concepts. In one study, Dutch and Farsi speakers were found to perform differently on a nonlinguistic task requiring them to reproduce musical pitches, in line with the spatial metaphors employed in their respective languages. The researchers concluded that "speakers of different languages tend to form systematically different representations of the same physical experiences, even when they are not using language" (Dolscheid, Shayan, Majid, et al.

2013, 620). This claim may be questioned on principled grounds, since some researchers have doubted that effects of language use could somehow "reach down" and affect such abilities as perception and memory (e.g. Pinker 1995). But there is ample evidence of top-down effects in other areas of cognition and it is well established that language use can lead to changes in perceptual discrimination abilities. To take just one relevant example from phonology, very young infants initially discriminate consonant contrasts that are not made in their native languages, but by the age of ten to twelve months old, they no longer make these contrasts and perform similarly to the adults in their linguistic communities (Werker & Tees 1984; cf. Gleitman & Papafragou 2005). This is an effect of phonology, not syntax or semantics, on auditory perception, but it demonstrates that even relatively limited exposure to language can have an influence on perceptual discrimination abilities.[9] Other effects of language use on thought can proceed in a similar fashion by affecting preexisting innate predispositions. Language use may not create these dispositions but modify preexisting dispositions, either enhancing or diminishing them.

To summarize, there is considerable evidence to suggest that there are two types of process that result in what are generally taken to be LT effects. Since they operate on different time-scales and exploit different cognitive resources (e.g. covert labelling, habitual associations), it is reasonable to conclude that these processes mark out different causal pathways, and hence, correspond to different cognitive kinds. As mentioned earlier, different experimental paradigms have been developed to detect these processes, and these include verbal interference designs intended to disrupt the simultaneous and covert recruitment of linguistic labels during the performance of a nonlinguistic task, thereby distinguishing online processes from offline ones. These experimental paradigms have been used to support the hypothesis that there are two cognitive kinds at issue. Some researchers suggest that the ultimate vindication of this claim would lie in the discovery of neural mechanisms that are proprietary to each cognitive process. Neural mechanisms may yet be discovered that pertain to each of these processes, and while reduction of each cognitive process to a type of neural process cannot be ruled out, it should not be assumed either. Unlike some other cases analyzed in previous chapters, in the case of LT processes, there does not seem to be an obvious obstacle to the identification

---

[9] In the study by Dolscheid, Shayan, Majid, et al. (2013) on musical pitch reproduction in Dutch and Farsi speakers, they trained Dutch speakers to employ Farsi-like spatial metaphors for pitch, as a result of which Dutch speakers' performance resembled that of Farsi speakers. See also footnote 5, above.

of a type reduction of cognitive kinds to neural kinds. Even though LT effects concern the conceptual contents associated with linguistic terms, and I argued in Chapter 2 that conceptual content is individuated (in part) externalistically, what appears to be relevant in these cases is the internalist component of the conceptual contents. In other words, these cognitive effects pertain to language users' conceptions, not the public meanings of concepts in their languages or dialects. Hence, the two posited types of LT process in question may have types of neural counterparts.

These effects of language on thought can help illuminate the ways in which language does not merely act as a medium to convey and communicate thought, but enters into the structuring and shaping of thought itself. These structural effects on thought corroborate the claim made in Chapter 2 that language is not just a passive medium for the expression of concepts but rather enters into the very individuation of concepts. Moreover, this understanding of LT effects and processes may shed some light on what can be termed "general" LT effects. A weaker but more general thesis associated with language and thought is that language possession (as opposed to not having language at all) helps equip creatures with certain cognitive abilities that they otherwise would not have had.[10] The previous discussion of two kinds of LT process can provide some further understanding of how this is possible. The first type of "online" LT process is one whereby a language user deploys a label to help them perform a nonlinguistic task, such as a perceptual or memorial one, and this could explain why language users might have an advantage in performing some cognitive tasks that do not involve language. Cognitive agents who use language can avail themselves of this labelling device even when performing nonlinguistic tasks, involving perception, learning, and memory. The second type of "offline" LT process results from associations formed on the basis of extensive language use, and can also assist with certain cognitive tasks by strengthening associations among phenomena that are correlated in reality. If the existence of these processes is corroborated by further inquiry, then they could also explain why language users might have an edge over nonusers when it comes to their ability to engage in solving certain problems in the cognitive domain. Of course, the cognitive impact of being a language user need not be limited to these effects, but they may at least be part of the picture (see Figure 6.1 for a diagrammatic representation of the relationship to language-thought processes to processes of conceptual change and indicating their division into offline and online processes).

---

[10] See footnote 4 above.

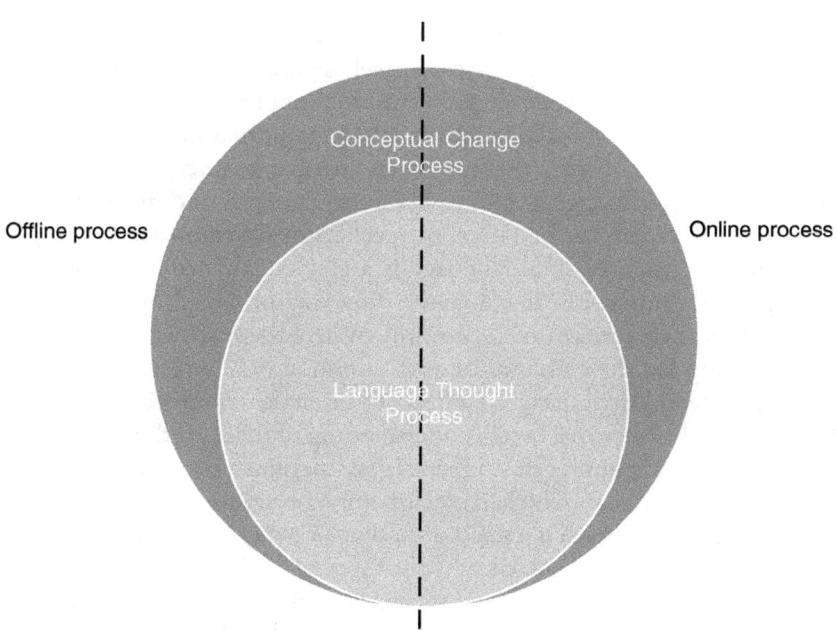

Figure 6.1. Language-thought processes are not a cognitive kind: schematic diagram showing the relation of language-thought processes to processes of conceptual change and indicating their division into offline and online processes.

6.6 Conclusion

Having tried to formulate the LT hypothesis with sufficient precision that it can be evaluated scientifically, it seems likely that it does not identify a single or distinctive cognitive kind. It is not distinctive because the processes that mediate the effects of language on cognition cannot be distinguished in a principled way from a broader class of processes (notably, the lexicalization of novel concepts within a natural language, leading to cognitive restructuring). Moreover, it is not a single kind because these processes do not seem to be homogeneous, since some of them involve covert and simultaneous recruitment of language while others do not, as outlined in the previous section. Perhaps the only thing that sets apart those phenomena regularly classified as LT effects is that they are unexpected and significant cognitive effects involving natural language. Cognitive differences that align with the (somewhat artificial) boundaries between natural languages are clearly of interest to human investigators, particularly since language is commonly thought to be a transparent medium that serves merely to convey thought rather than to shape or influence it in some way. But such effects may not be cognitively significant in their own right. Does this mean that the whole project of looking for LT effects ought to be abandoned? Not necessarily, but it does suggest that investigators need to look more broadly at cognitive effects involving conceptual misalignments among psychological subjects, whether speakers of different natural languages, or experts and novices, or speakers of different dialects within the same language. Natural language is a convenient marker of possible conceptual disparity, but it is not the only way in which human beings can differ in their conceptual repertoires. Where there are such disparities, there is room for the existence of both "online" and "offline" LT effects, as we have seen. Attributing LT effects to (at least) two distinct cognitive kinds opens the door to a better understanding of these kinds of processes. The natural-kinds approach confirms that there are two different kinds of causal pathways by which language might produce robust effects on human cognition, which is a realization that seems to have emerged gradually among researchers working in this area over the past decade or so. Also, by adopting a natural-kinds approach to the question of the interaction of language and thought, it becomes clearer that there is no principled distinction between LT processes and processes of conceptual change or augmentation.[11] This suggests that researchers need to inquire further into processes that involve cognitive restructuring, to determine the continuities and discontinuities between those involving competence in a natural language (e.g. mastery of color terms, spatial terms, gender terms) and those involving the acquisition of novel concepts in specialized domains, such as scientific theories domains of expertise (e.g. mastery of terms from surfing, wine-tasting, physics, climate science).

---

[11] Even though there may not be a strictly scientific reason for lumping all LT effects together and treating them as a separate class of cognitive phenomena, it is understandable that we might be especially interested in the cognitive effects that result from speaking different natural languages, for at least two reasons. First, as mentioned above, language is often considered a neutral medium for expressing thoughts rather than a shaper of those thoughts, and hence any evidence showing that natural language has some systematic influence on the thought of language users is of interest. That is particularly the case because the influence has sometimes gone unnoticed (and perhaps because those who have noticed it have tended to exaggerate it). Second, differences between natural languages are important to human beings given the cultural and political associations of natural languages. It is not surprising that any correlation between one's natural language and one's thought processes, albeit small, is of widespread interest both within and outside the scientific community.

CHAPTER 7

Cognitive Heuristics and Biases

(cowritten with Joshua Mugg)

> Man is not a rational animal, he is a rationalizing animal.
> – Robert Heinlein, *Tunnel in the Sky*
>
> I think unconscious bias is one of the hardest things to get at.
> – Ruth Bader Ginsburg

7.1 Introduction

This chapter focuses on heuristics and biases. A couple of preliminaries are in order. First, we are concerned with *cognitive* heuristics and biases rather than *social* or *implicit* bias. The phenomena we are interested in considering as examples of putative cognitive kinds involve competence in reasoning, inference, and decision-making. They concern cognitive tasks that deploy capacities for logical reasoning, inductive inference, probabilistic and statistical thinking, decision theory, and related norms of rational thought. Second, we take a *bias* to involve a deviation from the norms of rationality whereas a *heuristic* need not entail such a departure. We take the heuristic to be the underlying rule, process, or computation that may (or may not) result in a bias. This terminology agrees broadly with standard usage in cognitive science; for example, in an introduction to a seminal collection of papers, Gilovich and Griffin (2002, 3) define biases as "departures from the normative rational theory that served as markers or signatures of the underlying heuristics." Similarly, Stanovich, West, and Toplak (2016, 1110; original emphasis) write: "The term 'biases' refers to the systematic errors that people make in choosing actions and in estimating probabilities, and the term 'heuristic' refers to *why* people often make these errors – because they use mental shortcuts (heuristics) to solve many problems." Kahneman (2011, 7) says of participants in some of his experiments that "[t]he reliance on the heuristic caused predictable biases (systematic errors) in their predictions." Finally, even though he has a rather different position on the nature and prevalence of bias, Gigerenzer (2018, 306) states: "A bias is a systematic discrepancy between the (average) judgment of a person or a group and a true value or norm." Moreover, in his "fast and frugal heuristics" research program, heuristics are held to be "efficient cognitive processes that ignore information" (Gigerenzer & Brighton 2009, 107). Thus, there seems to be broad consensus in cognitive science that biases are systematic departures from norms of rationality, whereas heuristics are neutral rules that may or may not lead to bias, depending on the context of their deployment. In addition, for many of these researchers, heuristics are the underlying rules or principles from which cognitive biases stem. To be sure, some philosophers have used these terms somewhat differently. Antony (2016, 161; original emphasis) considers that "bias is an essential element in human epistemic success," and holds that "bias plays a *constructive* role in the development of human knowledge; it is an enabling condition of human cognitive achievement." In what follows, we will use the terminology that is more prevalent among cognitive scientists, not because we think that philosophers should always defer to scientific practice, but because it preserves a useful distinction between a pattern of reasoning that results in systematic error (bias) and one that does not necessarily do so (heuristic). In the rest of this chapter, our focus will be on heuristics rather than biases because we take heuristics to be more plausible candidates for kindhood than biases, as we shall explain, though we shall return to biases toward the end of the chapter. But before going on to focus on heuristics, we will outline two reasons for thinking that *cognitive bias* as an overarching category is not a promising candidate for being a real kind in cognitive science.

Cognitive biases seem to share a few common features. In addition to constituting departures or deviations from norms of rationality, they are also generally thought to be systematic errors in reasoning rather than occasional mistakes (as some of the above characterizations confirm). They are also considered to be widely shared among humans rather than idiosyncratic quirks found in a small number of individuals, though they are by no means always universal. Finally, it is often thought that they are hard to avoid, though many of them can be overcome with some instruction or by adopting debiasing strategies. But these additional commonalities (systematicity, prevalence, and relative unavoidability) are features found in many aspects of our psychological makeup and are not distinctive of cognitive bias (as opposed to, say, perceptual illusions), even when conjoined.[1]

---

[1] Additionally, although some cognitive biases are thought to be innate features of human cognition, many may not be innate but learned. Moreover, while some cognitive biases emerge more readily

7.1 Introduction

Hence, it would seem as though the only feature that sets cognitive biases apart is indeed that they are departures from rationality. But if this feature is to serve as a property common to all cognitive biases, we need to get clearer on the characterization of such departures. That obviously requires saying something about rationality. In the context of the cognitive science of reasoning and decision-making, rationality is thought to consist in a broad set of diverse norms, ranging from rules of logic, to rules of probabilistic and statistical inference, to those of decision theory. The deviations from the norms in these cases have very different natures: from those that constitute errors in logic to those that are considered to be errors in maximizing utility. Given the diversity that these departures represent and the different effects that they have on the actions and utterances of human agents, there does not seem to be much prospect for unifying them on a causal basis. Cognitive bias is unlikely to be a single kind of causal disposition or process with unified effects.

Moreover, the category of cognitive bias is almost certainly not a single etiological kind that has a common causal origin or history. Various researchers have speculated that biases generally have a variety of origins in the human mind, though there is considerable debate about the cognitive underpinnings of those origins. To mention just a few such proposed causal origins: lack of cognitive resources or "mindware gaps" (Stanovich, Toplak, & West 2008), "cognitive miserliness" (Toplak, West, & Stanovich 2011), motivational factors (Oreg & Bayazit 2009), or failure to inhibit intuitive responses (De Neys 2010). Though there is no consensus on which set of causal factors lead to cognitive biases, there is near unanimity among researchers that biases have multiple causes and can issue from various different aspects of the human psychological and cognitive makeup. Hence, there is no prospect of etiological unification when it comes to the category of cognitive bias.[2]

There is another problem with considering cognitive bias to be a cognitive kind. It follows from the characterization of biases and heuristics that we have adopted, which is prevalent in contemporary cognitive science, that heuristics are a more fundamental feature of human cognition or

---

when subjects are under cognitive load or under time constraints, others manifest even under optimal conditions.

[2] It may be useful here to invoke the analogy between cognitive biases and perceptual illusions, which has been deployed by numerous authors (cf. Kornblith 1994; Stein 1997; Kruglanski & Gigerenzer 2011; Pohl 2017). This analogy is apt in various ways, including the lack of a unifying causal basis in both cases. Even if we restrict ourselves to visual illusions, it is clear that they have multiple causal origins, ranging from basic physiological mechanisms to top-down interference from higher cognitive processes.

cognitive architecture than biases. While a biased response will be one possible effect of the heuristic, as we have seen, there may well be instances in which a heuristic manifests in veridical or rational responses as well. Thus, the heuristic will necessarily be more causally connected than the bias, since the bias is an effect of the heuristic, and the bias is related to underlying cognitive architecture via the heuristic. Therefore, we will proceed to examine the prospects for considering *cognitive heuristic* (or *heuristic* for short) to be a cognitive kind. After having examined the prospects for heuristics in Section 7.2, we will go on to look at the case of a more specific class of heuristics (*cognitive miserliness*) in Section 7.3. After concluding that neither the class of heuristics as a whole nor that particular subtype are good candidates for cognitive kinds, in Section 7.4, we will examine a yet more specific heuristic and its resultant bias (*myside* or *confirmation bias*), finding it to be a better candidate for a cognitive kind.

7.2 Heuristic as a Cognitive Kind

Although empirical work on heuristics and biases has proliferated for at least half a century and it now represents a significant and growing research program in cognitive science, there has not been much explicit attention devoted to the question whether heuristics constitute a cognitive kind. This question can be broken down into three separate questions:

a) Do all (or nearly all) heuristics collectively constitute a kind?
b) Do heuristics cluster in subtypes, such that one or more of these subtypes separately constitutes a kind?
c) Are there individual heuristics that are kinds?

We will consider these questions in this order, focusing on the first question in this section and the second and third in subsequent sections.

An early use of the term "heuristic" in cognitive science occurs in the classic paper by Newell and Simon (1976), "Computer Science as Empirical Inquiry: Symbols and Search," which is considered one of the founding documents of the field. There, Newell and Simon introduced the notion of a "Physical Symbol System" that "exercises its intelligence in problem solving by search" (1976, 120). The process that they called "heuristic search" is one whereby the system generates and progressively modifies symbol structures until it produces a solution to the cognitive problem. Crucially, the search is not an exhaustive one because such a system has limited processing resources; indeed, they emphasize that the resources are "scarce relative to the complexity of the situations with which they are

7.2 Heuristic as a Cognitive Kind

confronted" (Newell & Simon 1976, 120). Although this formulation is obviously vague, the key is that intelligent problem-solving always involves some selectivity rather than exhaustivity in the search for solutions. This proposed feature of heuristics, that they are *selective* cognitive processes, appears to be prevalent in many subsequent accounts in cognitive science. Moreover, a very common way to understand selectivity in this context is that it involves ignoring or omitting some information, as mentioned in Section 7.1. A closely related characterization is that heuristics solve problems by substituting a difficult problem with a simpler one that admits of an easier solution (Kahneman & Frederick 2002; Stanovich, Toplak, & West 2008, 263). The relation between the two formulations is not hard to find: One way of ignoring information is to substitute a complex problem with a simpler one, perhaps one that involves fewer variables or requires less processing. On this common understanding, heuristics are thought to be cognitive processes that perform well in certain contexts, despite the fact that they do not take into account all relevant information. This characterization would also seem to be consistent with some of the figurative characterizations of heuristics that are widely deployed, such as mental "shortcuts" or "rules-of-thumb," or colorful epithets that are applied to heuristics, such as "quick and dirty" (Gilovich & Griffin 2002) or "fast and frugal" (Gigerenzer 2004).

On its own, this characterization of a heuristic as *a cognitive process that ignores information* does not appear sufficient to establish it as a cognitive kind. We would need to know something further about the causal profile of such processes in order to determine whether this central feature of heuristics is either due to a common causal mechanism or issues in certain stable effects. But the prospects on either count are not promising. To see this, we will look at a couple of theoretical accounts of heuristics in cognitive science. The two most prominent rival accounts of the nature of heuristics are the one associated with dual-system theory and that posited by the research program of ecological rationality. In a "dual-system" or "dual-process" account of cognitive architecture, heuristics are generally considered to pertain to System 1, a collection of cognitive module-like systems that are supposed to have some common characteristics.[3] On this view, or family of views, there are two types of systems or processes that underlie human reasoning.

---

[3] Though there are substantive differences between dual-system and dual-process models, for our purposes here the differences do not matter greatly. On the former view, heuristics can be thought of as processes that are implemented by the systems, while on the latter view they are identical with such processes or at least some of them. In this section, we put things mainly in terms of dual systems, but what we say can be rephrased in terms of dual processes.

Broadly speaking, System 1 (*S1*) is fast, automatic, and associative, while System 2 (*S2*) is slow, controlled, and rule based. On early versions of these theories, heuristics were thought to pertain exclusively to *S1* and were held to be the default processes of human reasoning, which can be overridden or corrected by processes issuing from *S2* (e.g. Evans 1989). On many such views, heuristics from *S1* often yield valid responses when it comes to problems that the human mind has been adapted to solve. But in contexts that are far removed from the adaptive environment, they can give rise to inaccurate or mistaken solutions to problems. In such contexts, they need to be overridden by inferential or decision-making processes from *S2*, so as not to lead us into error or generate biases. It would seem that on this view there is a cognitive commonality to all heuristics, namely that they all issue from *S1*. But that is not generally agreed by dual system theorists themselves. For example, in an early articulation of the view, Evans (1989) labeled the two systems as "heuristic" and "analytic," respectively, but in later incarnations of the theory, he emphasized that heuristics can be associated with *S2* as well as *S1*. Later on, Evans (2011, 93) observed that "cognitive biases are as often attributed to Type 2 as Type 1 processing," pointing out that heuristic processing may occur in both systems and that cognitive biases may stem from both (cf. Evans 2012).[4] This would also seem to be the view of Tversky and Kahneman (1974) and most researchers employing the dual-system approach. But even if we were to narrow down the category of heuristics by applying it only to rule-based cognitive processing in *S1* (as some early versions of dual-system theory did), we would need to identify what all *S1* processes have in common, and that is in turn a vexed question without a clear or settled answer. In fact, a strong case has been made that the sub systems or processes of *S1* are not unified in terms of their causal properties. Despite the fact that many dual-system theorists once proposed a "Standard Menu" of features shared by all *S1* sub systems or processes, these early accounts have largely been abandoned in the face of significant difficulties. Proponents and critics alike have pointed out that the properties or features characteristic of each of *S1* and *S2* actually crosscut one another rather than cluster in certain ways (see e.g. Samuels 2009b; Evans & Stanovich 2013; Mugg 2016). Thus, even if heuristics are understood as *S1* processes they do not seem to share certain distinct causal properties that would set them apart from other cognitive processes and issue in certain stable effects.

---

[4] Evans (2012, 16) writes: "But it is an error to think that Type 1 processing is necessarily biased or that Type 2 processing is necessarily logical and abstract ... Both types of processing can lead to correct answers and both can lead to biases." Also: "with experience we may adopt quick and dirty heuristics which are still explicitly applied by Type 2 processing ..." (Evans 2012, 23)

Meanwhile, on ecological rationality views, heuristics are "fast and frugal" processes that are designed to solve various adaptive problems (see e.g. Gigerenzer & Brighton 2009; Gigerenzer 2018). On such theories, heuristics are sometimes superior to algorithms that take into account all available information. It is not just that heuristics involve a trade-off between accuracy and efficiency, since in some environments "heuristics are more accurate than strategies that use more information and computation" (Gigerenzer & Brighton 2009, 116). Gigerenzer and colleagues provide a number of examples to support this seemingly paradoxical claim. Consider something like the recognition heuristic, according to which thinkers choose between two alternatives based on their recognition of one of the alternatives. To illustrate, if American students are given pairs of German cities and asked to choose the most populous one in each pair, they typically choose the one whose name they recognize, since they generally lack detailed knowledge about the cities. The recognition heuristic turns out to be highly successful in this task, for the simple reason that size and recognition are highly correlated, at least for American students and German cities. In general, the heuristic works if there is a correlation between recognition and criterion in the environment (Kruglanski & Gigerenzer 2011, 100). Therefore, proponents of ecological rationality tend to regard heuristics as being more efficient and less prone to error or bias than dual-system theorists. But they do not claim that heuristics are always or even predominantly efficient or error-free. Moreover, from this perspective, heuristics can be either intuitive or deliberate cognitive processes; indeed the very same heuristic can be deployed intuitively or deliberately (Kruglanski & Gigerenzer 2011). To be sure, Gigerenzer (2004, 63–64) characterizes heuristics as having three features in common: (1) They exploit evolved capacities, (2) they exploit structures of environments, and (3) they are distinct from optimization models. But the first two features are clearly not distinctive of heuristics since many, if not most, aspects of cognition exploit evolved capacities and the environment. As for the third feature, this follows directly from the fact that heuristics do not take into account the totality of information in solving problems.[5] Hence, it is safe to say that heuristics, on this ecological rationality view, have nothing unique in common apart from the fact that they are cognitive processes that ignore information. If we classify a cognitive process as a heuristic there is nothing more we can say

---

[5] This is not an objection to Gigerenzer, since he seems to put these forward as necessary conditions on heuristics rather than properties that are causally related to the central characteristic of heuristics.

about it; there are no generalizations to be made beyond the one that we used to identify them in the first place.

This brief survey suggests that there is nothing common to cognitive processes that deploy or implement heuristics, beyond the fact that they ignore information, on either of the two major theoretical approaches to heuristics in cognitive science. In the following section, we will look at some recent attempts to further subdivide the category of heuristics into narrower categories, in order to determine whether there may be candidates for cognitive kinds among the subordinate categories of heuristics. In so doing, we will also further corroborate the claim that the category of heuristics itself does not seem to correspond to a cognitive kind.

7.3 Sub Categories of Heuristics as Cognitive Kinds

In the previous section, we considered whether the category *heuristic* corresponds to a cognitive kind. We concluded that, at least according to the dominant theoretical accounts of heuristics in the empirical literature, there was no feature that all heuristics shared beyond their being selective cognitive processes that do not take into account all relevant information. The question we need to consider in this section is whether any sub categories of heuristics might correspond to a cognitive kind. By focusing on what we take to be one of the most promising candidates, we will argue for an answer in the negative, at present, though we will conclude with a positive suggestion for researchers.

There is a certain family resemblance among some heuristics, such as vividness effects (e.g. representativeness bias), affect substitution, impulsively associative thinking, framing effects, anchoring effect, belief bias, denominator neglect, outcome bias, hindsight bias (also known as "curse of knowledge effects"), conjunction errors, and confirmation bias.[6] In fact, some look like instances, determinates, or subgroups of the others. For example, the anchoring effect looks like a framing effect pertaining to estimation of numbers when numbers are mentioned previously to the question. Likewise outcome bias, hindsight bias, belief bias, and confirmation bias look quite similar. There are, however, two problems with classifying distinct sub categories of heuristics by way of such family resemblance. First, as we shall see, there have been multiple attempts to categorize heuristics into subkinds, and they have not tended to match up very well. It seems that when we proceed by

---

[6] Although we label some of these as "biases" here, we do so only because that is what they are so called in the empirical literature. Again, we are interested in the heuristic underlying these biases.

7.3 Sub Categories of Heuristics as Cognitive Kinds

way of grouping heuristics together by mere similarity relations, there are too many ways to cut the cake, and several of the heuristics will fall into multiple categories. Second, a mere family resemblance is not enough for kindhood. In keeping with the account of cognitive kinds adopted in this book, we would need to understand *why* there is a family resemblance by establishing the *causal* connectedness of the shared properties. As of yet, such a principled causal basis for subdividing heuristics has yet to be provided.

This last claim can be further supported by taking a look at some proposed taxonomies of heuristics. We will argue that the nature of these categories also provides indirect support for the claim that there is no category, *heuristic*, characterized by a set of common features, that can be divided into subtypes that relate to it as species to genus. These taxonomies attempt to group heuristics and biases into a number of clusters based on their differentiating features. There have been a number of attempts to divide the domain of heuristics (and biases) into taxonomic categories, based on a variety of principles or theoretical considerations, but these sub categories are often disparate in terms of their causal and etiological profiles. Ceschi, Costantini, Sartori, et al. (2019) recently undertook an attempt to compare some of the prominent taxonomies in the cognitive science literature and tabulated the results for ease of comparison (see Figure 7.1).

There are several things to notice about this attempt to compare various taxonomies. First, most of these taxonomies do not distinguish clearly between heuristics and biases – though we might interpret this charitably to mean that they are interested in the heuristic underlying the bias in all cases. Second, different taxonomies deploy divergent categories to classify heuristics and biases and there is almost no overlap in the labels that they give to the categories that are deployed. Of course it may be that some categories as deployed by some theorists are just terminological variants of those adopted by other theorists (e.g. in Figure 7.1. Arnott's "Adjustment" category may correspond to Carter, Kaufmann, and Michel's "Reference Point" category, and Stanovich et al.'s "mindware gaps" may correspond to Oreg and Bayazit's "simplification biases"). But a little probing suggests that their categories only partially overlap and may even crosscut (e.g. Baron's "representativeness" and "availability" categories overlap Oreg and Bayazit's "simplification biases," which also includes phenomena not classified by Baron).[7] Third, and most importantly, even within each taxonomy, the categories are

---

[7] While there is nothing wrong with crosscutting categories in science (see Khalidi 2013), in this case, it does not appear that the different taxonomies are trying to capture different aspects of the heuristics (as when biologists classify organisms based on phylogenetic and ecological properties).

Source / Heuristic / Bias	Baron (2000)	Gilovich, Griffin, & Kahneman (2002)	Arnott (2006)	Carter, Kaufmann, & Michel (2007)	Stanovich, Toplak, & West (2008)	Oreg & Bayazit (2009)
Gambler's fallacy	Representativeness	Representativeness and Availability	Statistical	Control illusion	Probability knowledge (Mindware gaps)	Simplification biases
Conjunction fallacy	Representativeness	Representativeness and Availability	Statistical	Control illusion	Probability knowledge (Mindware gaps)	Simplification biases
Representativeness heuristic	Representativeness	Representativeness and Availability	Statistical	Base rate	Probability knowledge (Mindware gaps)	Simplification biases
Base rate fallacy		Representativeness and Availability	Statistical	Base rate	Probability knowledge (Mindware gaps)	Simplification biases
Framing			Presentation	Presentation	Focal Bias	Regulation biases
Distinction bias						
Availability heuristic	Availability	Representativeness and Availability		Availability cognition	Mindware gaps	Simplification biases
Imaginability bias	Availability	Representativeness and Availability	Memory	Availability cognition	Mindware gaps	Simplification biases
Better than average effect				Output evaluation		Verification biases

Optimism bias	Effect of desire on belief	Optimism	Confidence	Output evaluation		Verification biases
Anchoring heuristic	Underadjustment	Anchoring contamination and Compatibility	Adjustment	Reference point	Focal bias	
Reference price			Adjustment	Reference point		
Regression toward the mean			Adjustment	Reference point		
Extra cost effect	Utility theory					
Sunk costs fallacy	Utility theory			Commitment		
Endowment effect	Diminishing sensitivity			Commitment		Regulation biases
Time discounting	Diminishing sensitivity					

Figure 7.1. Comparison of taxonomies of heuristics and biases from five different sources; parentheses indicates classification in more than one category (adapted from Ceschi, Costantini, Sartori, et al. 2019).

based on rather diverse theoretical considerations. This last point is perhaps the most significant for our purposes since it signals that there does not seem to be a common basis for classification even by a single group of theorists. For example, Oreg and Bayazit (2009) draw a tripartite distinction among biases based on the motivations that give rise to the biases (simplification, verification, regulation). Roughly speaking, simplification biases stem from a desire to achieve a comprehensible image of the world, while verification biases are motivated by the need to achieve consistency and coherence, and regulation biases arise from trying to approach pleasure and avoid pain. But they explicitly argue that there are complex direct and indirect relationships among the bias categories. For example, according to them, "verification biases contribute to the creation of regulation biases" (Oreg & Bayazit 2009, 189). Hence, the underlying dispositions are not independent of one another and do not divide the biases into disjoint categories. Similarly, Stanovich, Toplak, and West (2008) provide a taxonomy of heuristics and biases that is based on the nature of the breakdown in reasoning that results in the bias (e.g. cognitive miserliness, mindware gap, contaminated mindware), but they also acknowledge that some biases may belong to more than one category. Setting aside the fact that they lump heuristics and biases together, what this suggests is that some of the heuristics that they have identified are due to some general features of cognitive processing (e.g. cognitive miserliness) while others are just a result of a lack of cognitive skill or training (e.g. mindware gap). Finally, with regards to the kindhood of the superordinate category *heuristic*, none of these taxonomies identifies any common traits of heuristics – beyond ignoring information – that would unify them and enable us to distinguish their sub categories in the manner of genus and differentia. When one looks at the categories within each taxonomy, let alone across taxonomies, they do not seem to have any obvious shared features that would identify them all as subordinate categories of the superordinate category, *heuristic*. Thus, these taxonomies also provide indirect support for the claim that *heuristic* is not a homogeneous category, since its subordinate categories are not characterized by a common set of properties (beyond ignoring information).

This last conclusion is further corroborated by what seems to be the most comprehensive and "evidence-based" approach to establish a taxonomy of heuristics and biases. In arriving at their own taxonomy, Ceschi, Costantini, Sartori, et al. (2019) build partly on existing taxonomies (tabulated above) but they also pursue a strategy that relies on finding correlations between the performance of subjects on different cognitive tasks associated with heuristics and biases. Using a large number of participants

7.3 Sub Categories of Heuristics as Cognitive Kinds

(n = 289) and a within-subjects design, they attempt to discern patterns of correlation in performance on a battery of seventeen cognitive tasks. Their method involves complex statistical techniques for discerning patterns among these tasks, but the method can be divided into two main steps. First, they performed a Multiple Correspondence Analysis (MCA) to determine the presence of common categories deployed in existing taxonomies. Then they performed a Principle Component Analysis (PCA) to assess relationships between biases belonging to the dimensions extracted from the MCA. This yielded three factors that were interpreted as the main categorical distinctions between types of heuristics or biases: (1) mindware gaps, (2) valuation biases, and (3) anchoring and adjusting. Here again, there is no suggestion that all heuristics have commonalities beyond ignoring information. The subordinate categories are not identified as species of a genus, each of which is characterized by a number of common properties in addition to certain distinguishing characteristics that differentiate them from other members of the genus. These considerations confirm the heterogeneous nature of the category of heuristics. But although this taxonomy, like the others already considered, does not give us any reasons for thinking that *heuristic* is a valid cognitive kind, it is possible that some of these subordinate categories correspond to cognitive kinds, not as species of a single genus but as stand-alone kinds. The specific categories that Ceschi, Costantini, Sartori, et al. (2019) identify do not seem to be promising candidates, as can be surmised by considering each very briefly. *Mindware gaps*, posited by Stanovich (2010), are instances in which thinkers simply lack the relevant reasoning principles to perform a cognitive task, as when they commit the conjunction fallacy, gambler's fallacy, or base rate fallacy. These supposed gaps in knowledge are clearly cognitively variegated and have nothing in common apart from being instances of ignorance of certain rules or principles. As for *valuation biases*, they include such biases as optimism bias, temporal discounting, and sunk cost fallacy, and appear to involve either over- or under-valuing certain outcomes. As Ceschi, Costantini, Sartori, et al. (2019, 197) admit, those who are susceptible to such biases can have opposing character traits (e.g. optimism and pessimism), which also suggests heterogeneity of the bias. Similarly, *anchoring and adjustment*, which includes framing effects and regression to the mean, seems to stem from a variety of traits or dispositions and is not correlated with other cognitive features. We will therefore take a closer look at another prominent sub category of heuristics, which is identified with its role in a specific cognitive architecture, to determine whether it might correspond to a cognitive kind in its own right.

If cognitive kinds are individuated in terms of their causal role, a promising approach to distinguishing a sub category of heuristics is to do so with reference to a causal model of cognitive architecture. We have argued that heuristics all involve ignoring information or taking some sort of short-cut, but it is plausible that not all information ignoring or short-cutting are the same. It may be possible to identify various points in cognitive processing where information is ignored and identify corresponding heuristics. We might find patterns among the various individual heuristics based on the point at which information is ignored within cognitive processing. This would provide us with a taxonomy of heuristics, subdividing the category *heuristic* into genuine kinds, with each sub-kind having a distinct causal profile in human cognitive architecture. Such a taxonomy would provide us with an explanation for why the heuristics are so divided by relating each subdivision to the causal structure of cognitive processing. It would also aid in cases where it seems a heuristic fits into more than two categories: When specific heuristics seem to fit into more than one of these categories, we would have good reason to split the heuristic, as it would be implicated at two distinct points in reasoning processing. Some of the taxonomies of heuristics already mentioned seem to be attempting to divide heuristics along these lines. For example, Stanovich offers a framework for conceptualizing individual difference, which fits naturally with his taxonomy for classifying heuristics. This framework can provide the basis for identifying certain subtypes of heuristics that are likely to be candidates for cognitive kinds. The plausibility of sub-groupings of heuristics being kinds depends upon which category we consider and the cognitive architecture in which it is situated. It is not possible to run through every sub category that has been mentioned in the literature, but we can at least consider one of the more promising sub categories proposed in one of the taxonomies already mentioned. Stanovich claims that there is a group of heuristics that are unified because they all arise from *cognitive miserliness,* which itself arises, in part, because of time and computational restraints. In the remainder of this section, we will consider whether *cognitive miserliness,* as proposed by Stanovich, is a cognitive kind capable of unifying some of the items in the above list of individual heuristics.

In an attempt to describe cognitive miserliness, Stanovich, Toplak, and West (2008) break it down into two aspects or "rules." The first rule is: "Default to Type 1 processing whenever possible." This rule is conceived as a structural feature of our cognitive architecture and it presupposes a dual-process view of cognition. As such, its fortunes are tied to those of dual-process theory: If it turns out that dual-process theory is not an apt

7.3 Sub Categories of Heuristics as Cognitive Kinds

theoretical account of human cognition, then this first aspect of cognitive miserliness simply cannot be sustained in its current form. Setting aside worries about dual process theory outlined in Section 7.2, let us grant for the sake of argument that something like that theory is correct. If we accept this cognitive architecture (or something like it), the construct of cognitive miserliness would seem to be a second-order rule or process that governs the Type 1 processes posited by dual-process theory. Cognitive miserliness is a process whereby cognition defaults to Type 1 processing. Presumably, a basic feature of cognitive architecture is that Type 1 is the default processing that conserves effort and requires fewer resources. But, according to Stanovich, Toplak, and West (2008), this is not the only way in which cognitive miserliness features in cognitive architecture. The second aspect or rule of cognitive miserliness is activated when Type 1 processing will not yield a solution; at that point, the thinker relies on serial associative cognition with a focal bias (which is a Type 2 heuristic process). Stanovich (2010, 67) expresses the "basic idea" behind focal bias as follows: "… the information processor is strongly disposed to deal only with the most easily constructed cognitive model." He also writes that "There are less expensive kinds of Type 2 processing that we tend to fall back on when Type 1 mechanisms are not available for solving the problem" (Stanovich 2010, 63). This takes place particularly in novel situations from an evolutionary point of view, where there are no stimuli that trigger Type 1 processes. Again, this is a second-order rule or process rather than a case of first-order cognitive processing.

This account of cognitive miserliness is couched in a cognitive architecture that posits three hierarchically organized systems in the mind: (1) the Autonomous Mind (associated with Type 1 processing); (2) the Algorithmic Mind (associated with Type 2 processing, responsible for cognitive ability measured by intelligence tests); and (3) the Reflective Mind (also associated with Type 2 processing, responsible for different cognitive styles) (Stanovich 2010, 35). This scheme is depicted in diagrammatic form in Figure 7.2. In this boxology, cognitive miserliness can be identified with two distinct processes. The first is represented by the horizontal arrow leading directly from the Autonomous Mind (responsible for Type 1 processing) to a response. The second is represented by the arrow labeled "E," leading from the Algorithmic Mind to response or attention. This suggests that cognitive miserliness is manifested in two entirely different cognitive processes or sub processes, pertaining to distinct psychological systems or capacities. In one case, it corresponds to a default to Type 1 processing, in another a resort to associative cognition (which is a Type 2

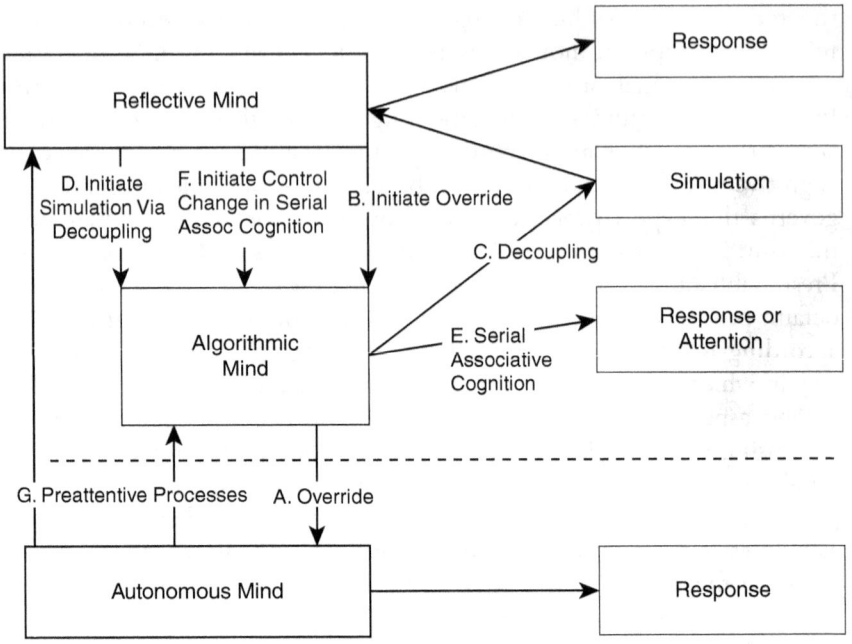

Figure 7.2. Tripartite model of the mind proposed by Stanovich (2010), showing the Autonomous Mind, Algorithmic Mind, and Reflective Mind, and some of their interactions.

process). As such, it does not seem to be a single type of cognitive process when viewed in the context of a causal model of cognitive architecture.[8] Now, it is possible that there is a unitary mechanism behind these different outcomes, perhaps some aspect of cognitive architecture that ensures that (in many contexts), cognition takes a path of least cognitive effort. Toplak, West, and Stanovich (2011, 1283) write: "Humans are cognitive misers because their basic tendency is to default to heuristic processing mechanisms of low computational expense." But if so, such an underlying cognitive tendency has yet to be described in any detail. Moreover, if it is a second-order rule that specifies when to use Type 1 and Type 2 processing, then it would seem to pertain not only to heuristics but to reasoning or cognition more generally.

[8] In fact, Stanovich sometimes mentions "override failure" as a "third category" of cognitive miserliness. He writes that "in override failure, cognitive decoupling does occur, but it fails to suppress the Type 1 processing of the autonomous mind" (Stanovich 2010, 100). This failure would seem to correspond to the absence of the arrows labeled "A" and "B" in Figure 7.2 in the performance of some cognitive task.

7.3 Sub Categories of Heuristics as Cognitive Kinds

In the face of this objection, defenders of cognitive miserliness might attempt to validate the construct by linking it to an operational test. Stanovich and collaborators have made the case that cognitive miserliness is subject to individual difference and is associated with a certain cognitive style. Rather than regarding it as a universal human trait that is uniform across individuals, they argue that empirical results show considerable variation among individuals in terms of their tendency to be cognitive misers. According to them, avoiding cognitive miserliness requires first detecting the inadequacy of the Type 1 response, then using Type 2 processing both to suppress the Type 1 response and to come up with a better alternative (Stanovich, Toplak, & West 2020, 1122; see also Figure 7.2 above). They hold that these abilities, which vary among individuals, are not measured on standard intelligence (IQ) tests, but are rather correlated with performance on the Cognitive Reflection Test (CRT), which consists of just three math questions that are fairly simple to solve, but also tempt experimental participants to offer an intuitive but incorrect answer.[9] Poor performance on the CRT is evidence of miserliness and good performance is evidence of the opposite. Unlike performance on standard intelligence tests or IQ tests, which is not perfectly correlated with performance on the full array of tasks in the heuristics and biases literature, performance on CRT is so correlated. It is, therefore, a more direct measure of cognitive miserliness. Toplak, West, and Stanovich (2011, 1284) put it thus:

> In short, the CRT is a measure of the tendency toward the class of reasoning error that derives from miserly processing. ... Intelligence tests do not assess the tendency toward miserly processing in the way that the CRT does. ... The CRT measures miserliness in action, so to speak. It is a direct measure of miserly processing rather than an indirect self-report indicator.

In subsequent work, Stanovich (2016) developed another, more extensive version of the test, the Comprehensive Assessment of Rational Thinking (CART), which is held to be a more accurate measure of cognitive miserliness. It might seem as though this would provide some corroboration of miserliness as a cognitive kind, since we have an instrument that is designed to measure it and assess the extent to which individuals are cognitive misers. But the existence of such a test is not a sufficient vindication of the existence of the kind. If we want to know what cognitive miserliness

---

[9] One of these is the notorious "bat-and-ball problem": A bat and ball together cost 10.10. The bat costs a dollar more than the ball. How much does the ball cost? The correct answer is: five cents (since 11.05 + 12.05 = 13), but many participants give the tempting answer: ten cents (see Kahneman & Frederick 2002).

is, we can say that it is whatever is measured by a certain test, but then if we want to know what that test measures, it seems that the only answer we have available is that it measures cognitive miserliness. The name of the revised test suggests that there is an independent construct being measured, namely Rational Thinking, but as pointed out in Section 7.1, rationality is a heterogeneous category. More importantly, rationality is far broader than just a lack of cognitive miserliness,[10] as Stanovich and colleagues agree, since they also think that rationality involves such attributes as avoiding "mindware gaps." The existence of the test is not sufficient to validate cognitive miserliness as a construct, nor establish it as a cognitive kind.[11] Therefore, *cognitive miserliness* does not unify a sub group of heuristics into a kind.

If Stanovich's attempt to categorize heuristics into subordinate kinds using *cognitive miserliness* fails, that does not imply that the overall approach of using cognitive architecture to provide a taxonomy of heuristics is flawed. There are other cognitive architectures, and each may be able to provide its own taxonomy of heuristics. Indeed, it may even be that a dual-process taxonomy of heuristics could be developed apart from cognitive miserliness. This provides a possibly fruitful avenue for researchers interested in looking at the relation between individual heuristics: to develop taxonomies of heuristics based on the various cognitive architectures currently on offer, with distinct sub categories of heuristics corresponding to elements of cognitive processing. One might worry about waiting on a completed cognitive architecture to determine the kindhood of the sub categories of heuristics. After all, the existence of the various

---

[10] In some work, the aim is said to be to come up with a test of the "Rationality Quotient" (RQ), along the lines of the Intelligent Quotient (IQ) measured by intelligence tests (Stanovich 2010, 189–190). Other constructs have also been proposed, such as "active open-minded thinking," but in the absence of some independent account of what these constructs are, this does not get us out of the circle.

[11] Could Stanovich and collaborators claim that one *part* of the test measures cognitive miserliness? Stanovich (2016, 29) identifies four subtests within CART that test for avoiding miserly processing, but he says the following about most of these subtests: "All of these tasks and their associated effects, although involving miserly processing, are still quite complex tasks. More than miserly processing is going on when someone answers suboptimally in all of them." So it does not seem as if any one subtest is an operational test for miserly processing (or the avoidance thereof). The one test that he implies is most geared to miserly processing is the "Reflection versus Intuition subtest" and the task that he mentions is the famous "bat-and-ball" problem (see footnote 8). However, he does not say whether we should take this task as an operational test for miserly processing. Even if we were to consider this task (and perhaps others like it) as an operationalization of the category of miserly cognitive processing, then that would not be sufficient to show that the construct is a valid one. We would still need a characterization of cognitive miserliness that situates it within a causal network. As famously argued by Cronbach and Meehl (1955): One needs to have a valid construct rooted in a "nomological net" before one can proceed to operationalize it (see also Flake & Fried 2020).

heuristics is the explanandum for which cognitive architectures are built as explanations. This may result in a temporary impasse, but it may not. With taxonomies of heuristics from the various cognitive architectures in hand, we can compare similarities and differences. One possibility is that the various cognitive architectures, though differing in where ignoring information figures within the overall cognitive processing, will produce similar taxonomies, indicating that some heuristics cluster to constitute cognitive kinds. In that case, we need not first determine which cognitive architecture is correct in order to determine whether some of the heuristics cluster into kinds. Of course, it may turn out that the taxonomies do not match up very well. In the meantime, we should consider whether individual heuristics are good candidates for kindhood.

7.4 Confirmation Bias or Myside Heuristic

Confirmation bias is one of the earliest biases discussed in the literature on heuristics and biases. In this section, we will begin by examining the evidence for a confirmation bias, but we will also consider the case for the existence of a *heuristic* underlying the bias, for reasons provided in Section 7.1. (In what follows, we will talk mainly in terms of "confirmation bias," since that is the preferred term in the empirical literature, but our real focus is the putative heuristic causing the bias.) Is there a cognitive kind corresponding to the category of *confirmation bias*, and if so, what are its main features? Moreover, what is the relationship between confirmation bias and myside bias, and should the former be replaced by the latter, as suggested by some researchers (e.g. Mercier 2017)?

Some of the earliest experiments that purported to show a confirmation bias in human subjects were reported by Wason (1960) using the so-called 2-4-6 task. The results that he obtained were supposed to show that a significant number of experimental participants are "unable, or unwilling, to test their hypotheses" (1960, 129). Wason gave participants the sequence of numbers 2-4-6 and asked them to guess the simple rule to which the numbers conform. To this end, participants were asked to provide guesses of other number triples to the experimenter and the reason for each guess, at which point the experimenter would tell them either that their guess was an instance of the rule or not. Participants were told that once they felt "highly confident" that they had hit upon the right rule, they were supposed to make an "announcement." The sequence 2-4-6 might suggest that the rule is something like "even numbers" or "consecutive even numbers" or "ascending consecutive even numbers," but the correct rule that Wason had in mind is just "ascending numbers." In the original experiment, six of twenty-nine participants (21 percent) guessed the correct rule on the first announcement, and ten (34 percent) guessed correctly on the second announcement. But the main finding was that those who did not get it right on the first or second announcement did not attempt to "test their hypotheses" in the sense of providing sequences that they did not think conformed to the rule, in order to rule out certain hypotheses. This has been taken to show that at least some people have a bias to confirm their hypotheses rather than disconfirm them (though Wason did not put it in these terms nor use the term "confirmation bias" in his original paper).

There are a number of things to notice about Wason's experiment and the conclusion that has often been drawn from it. First, a majority (55 percent) of participants performed quite well, hitting upon the correct rule on the first or second announcement and providing instances that were both compatible and incompatible with their hypotheses, thus effectively testing them (not just confirming them). Second, the task is a tricky one. As Wason (1960, 138) admits, one possible explanation of why some participants did not perform well (i.e. required more than two announcements) is that "the correct rule (increasing magnitude) was so trivial that students would have been reluctant to entertain it." Participants may not have provided enough negative instances because they did not consider any rules that are less specific than the obvious ones. Third, other psychologists have pointed out that this should not be seen as a decisive demonstration of a confirmation bias but rather a "positive test strategy," where a positive test strategy is one in which one tests cases that one thinks conform to one's hypothesis. In this case, if one's initial hypothesis is, "consecutive even numbers," then one would give instances conforming to it (e.g. 10-12-14, 98-100-102). In Wason's case, the true hypothesis was much broader than expected, so these guesses did not only conform to the participants' hypothesis, they were also in conformity with those of the experimenter. But if one's initial hypothesis is broader than the true hypothesis, or if it is overlapping, or disjoint, then proposing conforming instances can certainly lead to falsifying the hypothesis and suggesting alternatives (Klayman & Ha 1987; Klayman 1995). In general, adopting such a strategy need not produce systematic error, since one can discover that one's hypothesis is wrong by proceeding in this way, depending on the context. Hence, the experiment does not demonstrate a confirmation bias, though it may well show what could be labeled a "positive test heuristic."

In the decades since Wason's work, numerous researchers have pointed to cases that seem more like genuine instances of bias when it comes to confirming and disconfirming hypotheses. Edwards and Smith (1996) posit a "disconfirmation bias" when it comes to beliefs that are contrary to one's own beliefs. The emphasis in this work is on cases in which people attempt to *undermine* evidence that is contrary to their beliefs, though on a plausible model of credence, decreasing credence in an incompatible belief increases credence in one's existing beliefs.[12] In one experiment, Edwards and Smith (1996) chose seven issues about which participants had strong prior beliefs (as determined in pretesting several weeks preceding the experiment), such as the death penalty and corporal punishment. They presented participants with two arguments on each issue, consisting of a single premise and conclusion, one defending a certain position and the other defending the opposite position. In the first stage, participants were asked to rate the strength of each argument, and in the second stage, they were asked to list all the thoughts that occurred to them when they considered the conclusions of each of the arguments. They found that individuals judged arguments supporting beliefs that are incompatible with their own beliefs to be weaker, they spent more time scrutinizing the arguments, they generated a greater number of relevant thoughts about them, and they produced a greater number of arguments refuting those arguments (Edwards & Smith 1996, 14). This is often regarded as a seminal study showing that people are generally biased against arguments that conflict with or undermine their own beliefs. The bias consists both in a judgment concerning the strength of the opposing argument and in the time and effort expended in refuting it. Therefore, this can be considered a *disconfirmation* bias as opposed to a confirmation bias – but one directed at incompatible or contrary beliefs. Based on this and similar work, some researchers think that it is misleading to talk about a confirmation bias, since it is more accurate to say that people have a "myside bias" (Mercier 2017), favoring evidence that supports their own beliefs and disfavoring evidence that weakens them. Moreover, we would argue that since there are circumstances in which such a cognitive tendency may not be irrational or violate norms of inference (as we shall see shortly), it should be considered instead a "myside heuristic," given the terminology that we have adopted in this chapter. Therefore, in the rest of this section, we will address the question whether the *myside heuristic* can be considered a cognitive kind.

---

[12] Compare Mercier and Sperber (2011, 64) on confirmation bias: "It is a bias in favor of confirming one's own claims, which should be naturally complemented by a bias in favor of disconfirming opposing claims and counterarguments."

The main obstacle to considering *myside heuristic* to be a cognitive kind is the apparent heterogeneity when it comes to the kinds of psychological phenomena that it comprises. Some researchers have pointed out that the psychological processes that have been identified as factors in this body of experimental work range from relatively low-level perceptual or attentional mechanisms to higher-level cognitive dispositions to interpret and evaluate evidence and generate hypotheses. Many of these tendencies can be considered heuristics in the sense of rules or procedures that ignore information, but it may seem unlikely, given what we know about cognitive architecture, that it would be the very same process that is operative in these apparently disparate domains. Moreover, if these phenomena are all confirmed, it would seem that they often push in opposite directions, as it were, sometimes tending to confirm hypotheses and at other times tending to disconfirm them, depending on whether they are one's own hypotheses or incompatible ones. Could there be a unifying underlying cause that is responsible for all or at least a significant portion of these phenomena?

Given this apparent diversity, it would seem more promising to focus on one of the various phenomena at issue, such as the attitude toward evidence or arguments supporting and opposing one's own belief or hypothesis. If there is such an attitude, it would conform to the characterization of a heuristic that we're operating with in this chapter, since it is likely to involve one or more cognitive processes that ignore relevant information, in this case, either the actual evidence against a belief, the strength of that evidence, or the relative strength of the evidence for and against that belief. Nickerson (1998, 178) characterizes it as follows: "the tendency to give greater weight to information that is supportive of existing beliefs or opinions than to information that runs counter to them." What grounds do we have for positing a version of myside heuristic according to which subjects ignore arguments or evidence against their own beliefs in favor of evidence for those beliefs, and could such a tendency constitute a cognitive kind? Mercier and Sperber (2011, 63) hold that, as they have characterized it, confirmation (or myside) bias is "one of the most studied biases in psychology …" and survey some of the research studies in its favor. One such study has already been summarized above (Edwards & Smith, 1996) and it clearly illustrates that people judge evidence or arguments supporting their own beliefs to be stronger than evidence or arguments that are incompatible with their beliefs. In another influential study, Lord, Ross, and Lepper (1979) selected forty-eight participants, evenly divided among "proponents" and "opponents" of capital punishment, as determined in a pre test questionnaire. Participants were shown (fictitious) research results that either confirmed or disconfirmed their position, followed by detailed descriptions of the research procedure, along with critiques of the research and rebuttals by the supposed authors. All participants were exposed to both confirming and disconfirming information, counterbalanced to control for order effects. Asked for their final attitudes on the issue of capital punishment, relative to the experiment's start, proponents reported that they were more in favor of capital punishment, whereas opponents reported that they were less in favor (Lord, Ross, & Lepper 1979, 2103–2104). The researchers propose that when individuals encounter evidence that both supports and undermines one of their beliefs, they will assimilate the former while dismissing and discounting the latter. This "biased assimilation" of evidence in turn leads to belief polarization, whereby degrees of belief are strengthened rather than weakened after encountering both confirming and disconfirming evidence. Their data also support the existence of a "rebound effect," whereby participants are swayed temporarily by counter-evidence, only to revert to their former attitudes and beliefs or to even more extreme positions (Lord, Ross, & Lepper 1979, 2105). Improving on some of their methods,[13] Taber and Lodge (2006) claim to find stronger evidence for "belief polarization" when it comes to people's attitudes about highly charged political issues as gun control and affirmative action, especially when it comes to people with strong prior beliefs and those who are relatively sophisticated about the topics in question. One of the main causal factors that they identify as being responsible for this effect is that people with strong prior beliefs "evaluate supportive arguments as stronger and more compelling than opposing arguments" (Taber & Lodge 2006, 757). Similarly, Nyhan and Reifler (2010) found that members of ideological subgroups failed to revise their beliefs in the face of contrary evidence, and in some cases, strengthened their beliefs. In fact, they found a "backfire effect," whereby people maintained or strengthened their view even when confronted *only* with disconfirming evidence or arguments.[14]

---

[13] Taber and Lodge (2006, 756) critique the finding of belief polarization in Lord, Ross, and Lepper (1979), which is based on "subjective rather than direct measures of polarization," since they "asked subjects to report subjectively whether their attitudes had become more extreme after evaluating pro and con evidence on the efficacy of capital punishment." They claim to find evidence for belief polarization based on more objective measures.

[14] The terms "belief polarization" and "attitude polarization" tend to be used to denote strengthening or maintaining attitudes in the face of *both confirming and disconfirming* evidence. The terms "backfire effect" and "boomerang effect" tend to be used to denote strengthening or maintaining attitudes in the face of *only disconfirming* evidence. Some recent studies – for example, Wood and Porter 2019 – dispute the backfire effect, but they used simple factual statements by politicians (e.g. WMD were found in Iraq) that were then contradicted with factual corrections. Stanley, Henne,

They posit that this occurs because thinkers are motivated to come up with counter-arguments when they encounter disconfirming evidence, which just results in maintaining or strengthening their existing beliefs.

In sum, the tendency to maintain or strengthen beliefs when presented with evidence on both sides of an issue (or even just on the opposing side), has been widely attested in cognitive science, with a sizeable body of evidence to support it. Moreover, as we shall see in Chapter 8, some psychiatrists have posited this or a closely related cognitive disposition, "bias against disconfirmatory evidence" (BADE), in order to explain the emergence and persistence of delusions (Woodward, Moritz, Cuttler, et al. 2006). For some researchers, BADE in delusional patients is just the same tendency that exists in the general population,[15] but for others it is an accentuated or exaggerated form of a similar tendency in non-patients, and some evidence supports the view that delusional patients differ from controls in this respect (Woodward, Moritz, Cuttler, et al. 2006).[16] Hence, it may be a cognitive disposition that is present in a wide range of individuals to varying degrees. Alternatively, it may manifest itself in two varieties, one pathological and one non-pathological, with somewhat different characteristics. (This issue will be revisited in discussing psychiatric patients with delusions, specifically those diagnosed with Body Dysmorphic Disorder, in Chapter 8.)

The convergence of evidence from cognitive psychology, social psychology, political science, and psychiatry, including results obtained from a variety of experimental paradigms, suggests that a myside heuristic is widely, but perhaps not universally, manifested in the human cognitive makeup in a variety of contexts. Despite the use of different labels and taxonomic categories, a myside heuristic appears to be responsible for a variety of related effects, such as confirmation bias, disconfirmation bias, belief polarization, and the backfire effect, depending on the experimental

---

Yang, et al. (2020) also found no evidence of a backfire effect but they deliberately chose issues that are less contentious and emotionally charged (e.g. fracking, standardized testing) than those used in other studies (e.g. capital punishment, gun control).

[15] Maher (1988, 22) writes: "... deluded patients are like normal people – including scientists – who seem extremely resistant to giving up their preferred theories even in the face of damningly negative evidence" (cited in Woodward, Moritz, Cuttler, et al. 2006, 616).

[16] Woodward, Moritz, Cuttler, et al. (2006) devised a novel experimental test for identifying extreme cases of BADE, which involves showing participants three pictures comprising a story in reverse order, along with four possible verbal descriptions of the situation depicted. The descriptions that are most plausible given the first picture become less plausible as experimenters reveal the other two pictures, which show the same scene at earlier points in time. Delusional patients tend to stick to the initial description they selected (relative to controls), despite the disconfirming evidence.

condition in question. The empirical evidence suggests a tentative causal model for myside heuristic, along the following lines.[17] Thinkers who have a strong prior belief backed up by evidence or arguments encounter evidence or arguments that is incongruent with those beliefs. Such thinkers have an exaggerated confidence in their own initial belief and are motivated to defend it. They make an effort to refute the incongruent evidence, devising counter-arguments, finding flaws in the reasoning, reinterpreting it in such a way that it does not contradict their belief, or otherwise coming up with reasons to dismiss it. They go on to evaluate the incongruent evidence as being weak and this causes them not to assimilate or integrate it. In turn, this leads them to maintain or strengthen their confidence in their initial belief, now that they have refuted some (possibly new) counter-arguments or contrary evidence. This tentative sketch conceives of the myside heuristic as a causal process that pertains to our reasoning or inferential capacities, with some interaction between these inferential capacities and our motivations, along the lines of "hot cognition" (Kunda 1990).[18] Even though there may be other psychological factors that lead to similar results (e.g. perceptual, attentional, or memorial mechanisms), the primary one that we have been concerned with is an inferential process that is geared to evaluating arguments for and against a particular belief. Much of the empirical evidence points to a cognitive process that leads thinkers to evaluate arguments differently based on whether they are congruent or incongruent with their own beliefs. Under a variety of conditions, experimental participants evaluate arguments confirming their beliefs differently

---

[17] This sketch of a causal model draws on various sources. In describing the inferential process behind the confirmation bias, Klayman (1995) mentions such aspects as: overconfidence in one's initial belief, avoidance of performing tests that are likely to contradict one's hypothesis, interpreting evidence in such a way as to favor one's own hypothesis, insufficiently revising one's confidence in one's hypothesis based on contrary evidence, and reluctance to generate novel hypotheses in the face of new evidence. Edwards and Smith (1996) identify two phenomena at play: a judgment about the strength of the evidence and an effort expended to refute it. Nickerson (1998) posits two main factors: restriction of attention to a favored hypothesis and preferential treatment of evidence supporting existing beliefs. Taber and Lodge (2006, 757) say that the bias involves evaluation of arguments, differential time and resources devoted to arguing against incongruent as opposed to congruent arguments, and a preference for searching for confirming rather than disconfirming arguments. Stanovich, West, and Toplak (2013, 259) mention that the myside bias involves the generation of evidence, evaluation of evidence, and testing of hypotheses. Hahn and Harris (2014) say that confirmation bias is an umbrella term for a variety of ways that beliefs and expectations influence the selection, retention and evaluation of evidence, which overlaps significantly with "motivated reasoning," and they link it to research on "hot" cognition.

[18] In the context of research on psychiatric delusions, Bronstein and Cannon (2017) break down the bias against disconfirming evidence (BADE) into two factors, "Evidence Integration Impairment" and "Positive Response Bias," finding that the former but not the latter is associated with delusions.

from arguments disconfirming their beliefs, failing to revise their beliefs in the face of conflicting evidence, even strengthening their beliefs if they encounter both confirming and disconfirming evidence. There is clearly room for further research on the question of the causal network associated with the myside heuristic, particularly on the interaction between inferential and motivational processes and the possible involvement of other processes, such as perceptual, attentional, and memorial ones.[19] But we think that the cognitive process that we have sketched corresponds to a heuristic that treats evidence confirming and disconfirming one's hypotheses differentially, even though the proximal causes for such a heuristic are not fully understood.

One argument against the existence of a myside heuristic along the lines just delineated is that it would be maladaptive for humans to have such a disposition in their inferential toolkit, since it might seem to be irrational to be predisposed to treat evidence differentially depending on whether it is congruent or incongruent with one's own beliefs. The rational thing to do would surely be to treat all evidence in the same way and to follow it wherever it may point, regardless of prior beliefs. Indeed, it would seem to compromise the ability of human thinkers "to adapt effectively to changing environments" (Oswald & Grosjean 2004, 81; see also Nickerson 1998, 205–210). However, in at least some contexts and against certain background conditions, there are at least four ways in which a myside heuristic can be considered to be adaptive, in conformity with bounded or ecological rationality, or even in line with ideal theoretical norms. First, from the perspective of both ideal and bounded rationality, it is often rational to maintain one's hypothesis in the face of contrary evidence, particularly if that hypothesis has been strongly supported by past evidence and has survived other attempts at falsification or refutation (see e.g. Lord, Ross, & Lepper 1979, 2108; Nickerson 1998, 206–208; Taber & Lodge 2006, 767).[20] After all, both ideally and boundedly rational agents hold their beliefs for good reasons, so they ought not to abandon them lightly. Second, sticking to one's own beliefs in the face of countervailing evidence may lead to positive thoughts about one's judgment or opinions, generally resulting in self-affirmation (Munro & Stansbury 2009). It may be

---

[19] Rajsic, Wilson, and Pratt (2015) claim that there is a low-level perceptual mechanism biased toward confirmation. At some points they seem to be saying that this may reflect a general tendency toward confirmation in both perception and cognition, but at other times they indicate that there is just a similarity between the perceptual and cognitive processes.

[20] Some research finds a myside heuristic particularly or solely in sophisticated reasoners or those who hold strong opinions (e.g. Taber & Lodge 2006).

more comforting to cling to one's favored hypotheses, and in some cases this could have greater adaptive advantage than learning the truth about certain areas of interest (e.g. one's own abilities, the loyalty and affection of one's friends). Third, if one adopts the perspective of collective rationality and thinks of a community of thinkers along the lines of a debating society, it could be adaptive for each individual with a settled opinion to be committed strongly to that opinion, advocating for it in the face of contrary evidence, as long as a variety of hypotheses is entertained and each gets a fair shake. If such a debate is carried out for the benefit of the wider community, consisting largely of those who are not firmly convinced in any direction, and the community as a whole is allowed to decide on a course of action, this procedure may yield rational and adaptive outcomes. Something like this conception of the "marketplace of ideas" is widely thought to lead to rational decision making in the legal system and in the scientific community and is often considered to be "an efficient form of division of cognitive labor" (cf. Mercier & Sperber 2011, 65). Fourth, the adaptive advantage of a myside heuristic can also be defended if one holds that human inferential capacities have been selected for argumentation rather than reasoning (Mercier & Sperber 2011). In many situations, it may be adaptive to be persuasive, to "win friends and influence people," and persuaders who are wedded to their opinion and discount counterarguments may be more persuasive than those who are not. As Mercier and Sperber (2011, 63) put it, a confirmation bias "clearly serves the goal of convincing others." These four (not mutually exclusive) explanations of how a myside heuristic may be adaptive, and even rational, in certain contexts, shows that it might be a selected feature of our cognitive makeup rather than a dysfunction, and provides further reasons for thinking that it may be a cognitive kind. It also provides a possible etiology for the myside heuristic in terms of its distal causes.

Given that the myside heuristic can be seen to be adaptive, and indeed rational, does that mean that there is no myside *bias*, just a myside *heuristic*? Notwithstanding the arguments outlined in the previous paragraph, it is still possible that the myside heuristic may lead to systematic error in certain contexts, including some experimental contexts created in the lab. This means that there are grounds for identifying a myside bias as an offshoot of a myside heuristic. It bears emphasizing that instances of the myside heuristic that can be considered instances of a myside *bias* can only be identified against a broader background or context, including the particular task at hand, the social circumstances of the thinker, their degree of expertise, and the extent to which their own prior beliefs are justified.

This is another instance in which a cognitive kind can only be individuated relationally with reference to a particular environmental task or social context. Distinguishing a myside bias from its underlying heuristic is only possible relative to factors external to the thinker narrowly conceived.

7.5 Conclusion

For different reasons, the search for a unifying causal role of heuristics (in general) or biases (in general) is misguided. Biases are so designated because they are a systematic deviation from a rational norm, and given the heterogeneity of rationality, there is no reason to think that various rational errors will correspond to a homogeneous cognitive kind. There is no more reason to expect that cognitive biases constitute a kind than there are grounds for thinking that all visual illusions correspond to a kind. The category of cognitive heuristics that underlie these various biases also does not seem to be unified, since there are many ways in which the cognitive system can ignore information. It might seem, then, that the heuristics and biases research program, which purports to have discovered over 100 biases (and counting), rests on the mistaken idea that *cognitive bias* and *cognitive heuristic* are kinds.

Might we find patterns within the various heuristics and biases, allowing us to identify a subset of heuristics and biases as a kind? We have examined some recent attempts to provide taxonomies of heuristics, but they do not seem, at present, to point to any kind of consensus about how the various heuristics might cluster. It will not be enough to note a family resemblance relation among some subsets of the heuristics and biases. There should be something causally relating them to one another. We have identified Stanovich's *cognitive miserliness* as a putative kind unifying a subset of heuristics, as one of the more promising subtypes. However, upon closer examination *cognitive miserliness* lacks precision, and is not, as yet, a construct that corresponds to a real cognitive kind, or so we have argued. It may be that the ways in which subsets of heuristics and biases are grouped depends crucially on where and how information is ignored within the reasoning process. A taxonomy of heuristics will depend upon cognitive architecture, and as yet, there is no agreed upon cognitive architecture of human reasoning. As such, the prospects for identifying subgroupings of heuristics and biases that correspond to cognitive kinds may be grim at the present moment. Are we suggesting that the heuristics and biases research program is misguided? No, because when it comes to specific heuristics and biases, the picture is more promising. We have argued that there are at least good grounds for positing something like a *myside heuristic* or a *myside bias*, as an inferential cognitive process involved in evaluating and judging evidence for and against one's beliefs. Similar arguments might be made for some other individual heuristics and biases, though it is unlikely that every item in the menagerie of purported cognitive heuristics and biases (e.g. the IKEA effect or the Google effect), will turn out to be genuine kinds.

A final lesson that emerges from this examination of heuristics and biases concerns the distinction between a heuristic and a bias. We started by accepting the distinction often made by cognitive scientists between a bias, which is a systematic departure from rationality, and a heuristic, which is the underlying rule or process that sometimes eventuates in a bias. When it comes to the myside bias, in particular, this means that the bias can be distinguished from the heuristic only contextually, in relation to a particular task or problem, as well as a certain history of inquiry, and other factors. That is because discounting evidence against one's own favored hypothesis can be a bias in some contexts but not in others. This means that a bias is individuated both in relation to the environment of the thinker and the thinker's etiology. Therefore, if cognitive scientists have occasion to distinguish a myside bias from a myside heuristic, the former cognitive kind is externalistically individuated, and that is the basis on which it is distinguished from the latter kind. Hence, there is no prospect of identifying it with a particular neural process or structure, at least if these are individuated in the usual way in neuroscience, without reference to the broader environment or the history of the individual. Here again, we have an instance of a good candidate for a cognitive kind that is unlikely for this reason to be reducible to a neural kind.

CHAPTER 8

Body Dysmorphic Disorder

(cowritten with Amy MacKinnon)

> Taught from infancy that beauty is woman's sceptre, the mind shapes itself to the body, and roaming round its gilt cage, only seeks to adorn its prison.
> – Mary Wollstonecraft, *A Vindication of the Rights of Woman*

> All things counter, original, spare, strange;
> Whatever is fickle, freckled (who knows how?)
> With swift, slow; sweet, sour; adazzle, dim;
> He fathers-forth whose beauty is past change:
> Praise him.
> – Gerard Manley Hopkins, "Pied Beauty"

8.1 Introduction

A chapter on a psychiatric category may seem out of place in a book on cognitive ontology. But this chapter argues that the psychiatric category Body Dysmorphic Disorder (BDD) has a basis in perception and cognition. Specifically, we will propose a tentative causal model of the disorder that posits that it has certain key perceptual and cognitive deficits at its core, and argue that this causal profile makes BDD a strong candidate for being a real kind. Based in large part on this causal model, we will contend that BDD has been misclassified in the fifth and most recent edition of the *Diagnostic and Statistical Manual of Mental Disorders* (2013), DSM-5, which provides the most widely accepted taxonomy of psychiatric disorders. But before proceeding, we will try in this introductory section to respond to some common concerns about psychiatric taxonomy, specifically concerns that psychiatric categories cannot correspond to real kinds.

There has been growing discussion of psychiatric taxonomy in the philosophical literature, and there are a range of views on whether various different psychiatric categories correspond to real kinds (see e.g. Kincaid & Sullivan 2014; Pober 2013; Samuels 2009a; Tsou 2013; Weiskopf 2017a;

Ylikoski & Pöyhönen 2015). We will not be able to do these discussions justice here and will not attempt to situate our view in relation to them in any detail, but we will outline an account of psychiatric kinds and indicate some ways in which we diverge from some of the prevalent accounts of psychiatric disorders. While we will not try to make the case that the superordinate category *psychiatric disorder* corresponds to a kind, we will argue that BDD is a good candidate for being a psychiatric kind. As in some of the other cases discussed in this book (e.g. *memory* and *episodic memory*; *bias* and *myside bias*), we think it is possible for the subordinate category to correspond to a kind but not the superordinate category.

The view of psychiatric disorders that we are operating with considers them to be objective mental states or dispositions and ones that are individuated relative to social contexts. We are in broad agreement with the "harmful dysfunction" view of psychiatric disorders elaborated by Wakefield (1992) and other writers on psychiatric or mental disorders, with two important caveats. According to Wakefield (1992, 385), "A condition is a mental disorder if and only if (a) the condition causes some harm or deprivation of benefit to the person as judged by the standards of the person's culture (the value criterion), and (b) the condition results from the inability of some mental mechanism to perform its natural function, wherein a natural function is an effect that is part of the evolutionary explanation of the existence and structure of the mental mechanism (the explanatory criterion)." Although we agree that a mental disorder is a psychological malfunction of a certain type, we do not think the function at issue is exclusively a "proper function" in the evolutionary sense, and malfunctions are hence not necessarily biologically maladaptive (cf. Murphy 2006). Rather, malfunctions can be identified with breakdowns in systems or capacities that might not be associated strictly or directly with a biological adaptation of some kind. For example, some researchers have proposed that schizophrenia represents an evolutionary adaptation (for a review, see Polimeni & Reiss 2003). While the hypothesis may be implausible and has not had wide uptake, it does not seem possible to dismiss it on the grounds that psychiatric disorders are necessarily maladaptive. Functions in this context can be understood in terms of synchronic causal roles, and dysfunctions may involve disruptions in social interactions. The second caveat is that we do not think that identifying the harm involved in a psychiatric disorder involves a value judgment on the part of the clinician, as some writers appear to think. In this context, harm is understood in terms of systematic disruption to the ability of an individual to achieve their goals, avoid suffering, and establish meaningful relationships. This means that the psychiatrist who diagnoses a disorder need not endorse the values prevalent in the relevant social context that lead to the individual's being harmed. While the individual may suffer harm as a result of social pressures to conform or to fulfill certain duties, due to certain values held by those in the encompassing community, those values need not be shared by a researcher who is interested in categorizing the individual in question. We think that this shows that the researcher who categorizes an individual as having a psychiatric disorder is not thereby making a value judgment. A further point worth emphasizing is that the individual's abilities (to achieve their goals, avoid suffering, and establish meaningful relationships) are all exercised in a social setting, which means that they are relative to a social context. If an individual has certain psychological malfunctions that systematically and consistently disrupt these abilities and prevent their flourishing, they can be considered to suffer from a psychiatric disorder.

This account may invite the charge that it makes psychiatric conditions socially and culturally relative. But even though there may be some variability in what constitutes a psychiatric disorder on such a view, we do not think that this leads to strong a form of cultural relativism. The kinds of dysfunction that are at issue are ones that lead to disruptions in social relationships, vocational pursuits, and leisure activities, and they will apply across a wide range of human societies. Of course, it may be that some of the disorders listed in the DSM-5 do not qualify under this heading, but we would not regard that as a drawback for this view, since the DSM-5 cannot be considered the final word on what qualifies as a psychiatric disorder. It may also be the case that the current classification system of the DSM casts its net too wide, considering some conditions to be disorders that should not be so regarded. But when it comes to the particular disorder that we are discussing in this chapter, we think that it is clear that it conforms to the picture of psychiatric disorders that we have proposed, as we will try to show in due course.

A closely related concern is that this account of psychiatric disorders might be thought to give rise to what might be called the "drapetomania objection." The American physician Samuel Cartwright argued in the mid-nineteenth century that American slaves who tried to escape were afflicted with the compulsion to flee, labeling this psychiatric disorder "drapetomania" (see e.g. Wakefield 1992, 386; Murphy 2006, 27–28). How are we to avoid the conclusion that since this alleged condition arguably prevented its supposed sufferers from flourishing in their society at that historical juncture, it can rightly be considered a psychiatric disorder relative to that

8.1 Introduction

society at that time? The response is that genuine psychiatric disorders involve psychological dysfunctions, which can be assessed in terms of the individual's perceptual, cognitive, and affective abilities. Given that they involve psychological dysfunctions, psychiatric disorders do not just entail that those who suffer from them will violate local norms, but rather that they will issue in behaviors that are harmful to the individual in a broad range of social settings and result in thwarting their ability to achieve their goals. In particular, the harms that ensue cannot just be a result of labeling on the part of the society and the stigma attached to such labeling, since they issue from definite psychological dysfunctions.

If psychiatric disorders are identified with reference to the ability to function in a social setting, does that mean that *psychiatric disorder* is not a real kind, as characterized in Chapter 1 and subsequent chapters? We will not try to give a definitive answer to this question, but if it is a real kind, it is a relational kind, like many other kinds in both the basic and special sciences, and in this case the crucial relations are social ones. However, we would argue that individual psychiatric disorders may be real kinds, but not the superordinate category *psychiatric disorder*.[1] On this view, having a psychiatric disorder has real causes and effects, though both the causes and effects may be diverse (biological, psychological, and social).[2] This view of psychiatric disorders is generally in accord with the dominant "biopsychosocial model" that is widely adopted in psychiatry and the philosophy of psychiatry. Where we depart from this model is in doubting that each psychiatric disorder has a uniform biological cause (which may not currently be known).[3] In some cases, the account of a psychiatric disorder will be "psychosocial" rather than "biopsychosocial." Moreover, we think that the reason that psychiatric disorders are identified as such in the first place is because their effects are primarily of a psychological and social nature, involving social harm and dysfunction, though they may also have

---

[1] A simple analogy might help here. The category *dog* may correspond to a biological kind, even though the category *pet* does not. Dogs are pets and we may be particularly interested in them for that reason, but there need not be a real kind (whether biological or social) that includes all and only the species that we consider pets.

[2] On this point we agree with a number of philosophical accounts of psychiatric kinds, for example, Kornblith (1994/2014, 108–109) writes: "Consider, for example, debates in psychiatry about the proper characterization of various mental disorders. When disagreements arise about which symptoms or syndromes are to be classified together as falling under a single diagnostic category, what is at issue is not merely a matter of convenience, but rather a question that ultimately turns on the causal relations among the various alleged characteristics of the disorder." See also Pöyhönen (2011) for a view very similar to ours.

[3] It is not clear that the biopsychosocial model has that implication but that is how it is understood by some; for example, it seems to be the position adopted by Andreasen (1997).

biological causes and effects as well.[4] Hence, as in the case of several other cognitive kinds discussed in this book, it is not just that psychiatric disorders have social causes (among other causes), they are partly *individuated* in terms of social factors. The broader social context enters into their very identification as psychiatric disorders.

A prominent alternative to the view of psychiatric disorders that we have just outlined conceives of them primarily as dysfunctions of biological or neurophysiological systems (see e.g. Tsou 2016). But we would maintain that, even though there will be underlying neurophysiological bases for psychiatric disorders, these may be multiply realized, in accordance with the endorsement of multiple realizability for psychological kinds in earlier chapters. Also, in keeping with a theme sounded repeatedly in this book, the very same neural bases may not be correlated with a psychiatric disorder in different contexts. So the relationship between psychiatric disorders and neural states or dispositions may be many-to-many. For that reason, we would not expect to find that every psychiatric disorder can be identified with a single neurophysiological dysfunction, though this may be the case for some of them.

In this section, we have tried to address some common concerns about the status of psychiatric disorders as real kinds. Our arguments are certainly not decisive and this is not meant to be a full-blown defense of the real kind status of psychiatric disorders. In particular, there are wide variations in the types and levels of harm that ensue from psychiatric disorders, which may render them too heterogeneous to constitute a kind. A specific condition like BDD might be a kind of psychiatric condition with a unified causal profile even though it does not belong to an overarching kind that includes BDD and all (or even a significant number of) the other conditions that are commonly considered psychiatric disorders. In the rest of this chapter, we will begin by describing some of the main features of BDD, drawing on a considerable body of empirical and clinical evidence (Section 8.2). We will then go on to present some preliminary reasons for thinking that BDD has been misclassified in the DSM-5 (Section 8.3). We will then propose a tentative causal model of BDD, which emphasizes the internal states of individuals with BDD rather than simply their outward behaviors (Section 8.4). This will allow us to justify the claim that BDD is a real psychiatric kind. Finally, we will respond to some objections

---

[4] This claim would seem to distinguish the "biopsychosocial model" from the "medical model" of psychiatric disorders. Though we agree broadly with the account of psychiatric kinds developed in Murphy (2006), we disagree with him on this point.

concerning our causal model of BDD and our proposed reclassification (Section 8.5), before coming to a conclusion (Section 8.6).

8.2 Characterization of BDD

According to the DSM-5, BDD involves persistent and intrusive thoughts about a perceived bodily flaw that is not observable or appears slight to others. At some point during the course of the disorder, the individual will have performed repetitive behaviors (e.g. mirror checking, excessive grooming, skin picking, reassurance seeking) or mental acts (e.g. comparing his or her appearance with that of others) in response to concerns about appearance. The preoccupation causes clinically significant distress or impairment in social, occupational, or other areas of functioning. Moreover, the preoccupation with appearance is not better explained by concerns with body fat or weight, and symptoms do not otherwise meet diagnostic criteria for an eating disorder.[5]

There has been considerable empirical research, in both clinical and laboratory settings, on the main characteristics of BDD and its distinctive features. This work has revealed a number of important facets of the disorder that set it apart from other psychiatric conditions. In a seminal early monograph on BDD, Phillips (1996/2005) noted that neuropsychological studies suggest that those with BDD tend to over-focus on minor details when drawing complex figures from memory, compared to those without BDD. This has been corroborated by subsequent research suggesting that people with BDD are more likely to examine details in visual tasks and are less likely to take in the holistic picture. Their tendency to focus on details comes at the expense of processing global or configural aspects of a visual scene, and this is widely thought to be related to their tendency to concentrate on flaws in their own appearance. Feusner, Moller, Altstein, et al. (2010) tested this in the laboratory by comparing individuals with BDD with healthy controls on a task involving the identification of inverted faces. In healthy individuals, face recognition relies on both featural information (e.g. skin smoothness, blemishes, lines, hair texture) and configural information

---

[5] The last part of the definition may raise legitimate concerns about the reality of the condition, since it is defined at least partly in terms of what it is not. We will go on to argue that BDD should be characterized in terms of its underlying causal features, so the definition in the DSM-5 merely serves as the starting point of this inquiry into the nature of BDD. A number of instruments have been developed to help diagnose BDD, such as the Body Dysmorphic Disorder Questionnaire (BDDQ) (Phillips 1996/2005) and the Dysmorphic Concern Questionnaire (DCQ) (Oosthuizen, Lambert, & Castle 1998).

(e.g. spatial relationships of features, distances between features, holistic elements). Most experimental participants have been shown to demonstrate a "face inversion effect" whereby they are less accurate and slower at processing inverted faces, which requires greater reliance on featural rather than configural information (since some of those features change with inversion). If people with BDD generally rely more on featural rather than configural information, it was hypothesized that they may be expected to demonstrate less of a face inversion effect than controls, and this hypothesis has been supported by several studies, at least for reaction times though not for accuracy (e.g. Feusner, Moller, Altstein, et al. 2010). Feusner, Hembacher, Moller, et al. (2011) and Bohon, Hembacher, Moller, et al. (2012) confirm that individuals with BDD tend to focus on details at the expense of global aspects of images, especially when it comes to processing human faces and bodies. Moreover, Feusner, Hembacher, Moller, et al. (2011) found that the individuals with BDD "may have general abnormalities [relative to controls] in higher- and lower-order visual processing, beyond that for their own appearance or for faces in general." Using fMRI data, the same study reported that individuals with BDD showed greater activation in medial prefrontal regions for high spatial frequency images, indicating that they have "abnormal brain activation patterns when viewing objects," not just faces and bodies (Feusner, Hembacher, Moller, et al. 2011, 2385).[6] Some researchers have suggested that these visual processing abnormalities may play a role in the etiology of BDD. By showing that individuals who have not been diagnosed with BDD but are at risk of developing BDD (individuals with high Body Image Concern, BIC) exhibit these abnormalities in visual perception, these researchers hypothesize that the abnormalities may be part of what causes BDD, rather than a consequence of it (Beilharz, Atkins, Duncum, et al. 2016; see also Mundy & Sadusky 2014). They speculate that "abnormalities in visual perception mechanisms, specifically global and local processing, may be involved in the onset and maintenance of BDD ..." (Beilharz, Atkins, Duncum, et al. 2016; for a review see Beilharz, Castle, Grace, et al. 2017)

In addition to issues with visual processing, there is evidence that individuals with BDD differ from controls when it comes to their ability to recognize facial expressions and accurately interpret the thoughts of others

---

[6] Claims of direct correlations between abnormal brain activation (as measured by neuroimaging methods) and abnormal perceptual or cognitive processing need to be handled carefully. We would not set as much store by this evidence as these researchers appear to do and our argument does not take it as decisive.

(see e.g. Buhlmann, Wacker, Dziobek, et al. 2015). These deficits are subtle and do not emerge in all cognitive tasks involving the identification of mental states. Buhlmann, Winter, and Kathmann (2013) found that individuals with BDD did not differ from a control group with respect to their ability to interpret other people's emotional states based on images of their eyes, using the standard Reading the Mind in the Eyes Test (RMET), which has been widely used in studying theory of mind and social cognition. However, in a follow-up study, Buhlmann, Wacker, Dziobek, et al. (2015) found that individuals with BDD were less accurate than controls in correctly interpreting social situations when asked to evaluate scenarios depicted in video sequences showing interactions among four people. This task is more complex and dynamic than the RMET, since participants are instructed to evaluate the characters' emotions, thoughts, and intentions based on short videos rather than static images. Moreover, several studies have shown that people with BDD show a deficit in the recognition of emotional expressions in faces, especially when it comes to interpreting ambiguous expressions as negative or threatening and in "self-referent scenarios" (e.g. Buhlmann, Etcoff, & Wilhelm 2006; Grace, Toh, Buchanan, et al. 2019).

Individuals with BDD also differ from controls when it comes to the goals they set for themselves, at least as revealed in questionnaires and interviews. In one study, "internal goals" or the state of "feeling right" was found to be more important for those with BDD than for control participants (Baldock, Anson, & Veale 2012). Similarly, Veale and Riley (2001, 1390) report that "BDD patients are driven by a desire to camouflage their appearance or excessively groom to make themselves look their best or to feel 'comfortable.'" By contrast, controls are "motivated to use a mirror for more functional reasons such as making themselves look presentable or shaving." Prior to gazing, BDD patients are driven by the hope that they will look different, the desire to know exactly how they look, a belief that they will feel worse if they resist gazing, and the desire to camouflage themselves. After mirror-gazing, people with BDD often feel worse. No matter how many times they look in the mirror, they do not seem to reach a state of "goal completion" or "feeling right." Baldock, Anson, and Veale (2012) suggest that those with BDD are more likely to terminate mirror checking for reasons unrelated to their goals. For example, they do so because of frustration or time constraints rather than because of completing their goal and finally "feeling right" about their appearance.

Individuals with BDD are also characterized by delusions about their own appearance, which in some ways resemble delusions seen in psychosis

(Toh, Castle, & Mountjoy 2017). In the psychiatric literature, delusion and insight are often treated as opposite ends of a spectrum, and delusional beliefs are standardly characterized as "fixed beliefs that are not amenable to change in light of conflicting evidence" (DSM-5). Most accounts of delusions in both scientific and philosophical discussions no longer require that delusions be false, since some delusions may happen to be true, though the patient has no reason to believe in them. Hence, what is distinctive about delusions is thought to be, primarily, the extent to which the thinker resists giving up the belief, despite clear evidence to the contrary. Some researchers have distinguished the types of delusions that occur in BDD from those in schizophrenia and some other types of psychiatric disorders. According to Rosen (1995, 147), BDD delusions are not "bizarre," in the sense that (for example) "the patient might be preoccupied with 'ugly' vascular marks on the skin, but not with marks of the Devil." Accordingly, Rosen (1995, 148) suggests that "overvalued ideas" in BDD fall somewhere between obsessions and delusions in terms of insight, in that "the belief is entrenched and sensible to the patient, but he or she can acknowledge the possibility that it may not be true." But even though the beliefs associated with BDD may not be on a par with the delusions of schizophrenia, they clearly meet widely accepted criteria for delusions when it comes to their imperviousness to contrary evidence.

In this section, we have surveyed the evidence that shows how BDD is associated with certain significant cognitive deficits, regarding visual processing of faces, interpreting the emotions of others, and harboring certain delusions. In the next section, we will argue that these features and others provide a contrast between BDD and OCD, a disorder that it is classified with in the DSM-5.

8.3 Comparison of BDD with OCD

According to the DSM-5, the category of OCRDs includes five psychiatric disorders: OCD, BDD, trichotillomania, excoriation, and hoarding disorder. We will focus here on the similarities and differences among OCD and BDD, setting aside for these purposes the other three disorders. We take it that demonstrating a wide gap between BDD and OCD and important differences among their central features at least casts doubt on the plausibility of classifying BDD alongside OCD as one of the OCRDs. Others have commented generally on "intracategorical heterogeneity" in psychiatric classification (Held 2017), and some have even questioned whether BDD ought to be included within the OCRD category (for a review see e.g. Toh, Rossell, & Castle 2009), but we think that our reasons for questioning the classification are somewhat novel and shed further light on the nature of BDD. In what follows we will take our cue from the patient's perspective, particularly their mental states, in driving a wedge between BDD and OCD.

Though BDD and OCD both involve impulse control issues, there are some important differences between them. By trying to understand the patient's perspective and focusing on the patient's internal mental states as well as their outward behaviors, one can make a distinction between the emphasis on the *content* of the disorder (which occurs in BDD) and the emphasis on the *form* of the disorder (which occurs in OCD). This distinction tells us something important about the nature of these disorders. People with BDD are more focused on their image or particular body part (content), while people with OCD are more concerned with the intrusive preoccupations or thoughts themselves (form) (Rivera & Borda 2001).

Phillips, Pinto, Hart, et al. (2012) point out that insight/delusionality is a central construct of psychopathology, and that it is of increasing interest particularly within the area of OCRDs. As mentioned earlier, although BDD delusions are often not considered "bizarre," they can still be believed with such great conviction that they seriously impact people's ability to function. Phillips, Pinto, Hart, et al. (2012) studied individuals primarily diagnosed with OCD and those primarily with BDD, and assessed their levels of insight using the Brown Assessment of Beliefs Scale (BABS). This instrument was developed to assess the degree of insight and delusionality in patients with various psychiatric disorders, and it rates beliefs on various dimensions to emerge with a rating that ranges from good insight to no insight (delusionality).[7] When assessed using this measure, there was a wide range in the levels of insight within both OCD and BDD. But the distribution of scores differed significantly between the two disorders, with the majority of OCD participants showing excellent or good insight and the majority of BDD participants showing poor or absent insight. Another study found significantly reduced overall insight among BDD patients, 39 percent of whom were classified as delusional compared to only 2 percent of OCD patients (Eisen, Phillips, Coles, et al.

---

[7] The BABS scale requires an interviewer to identify one of the patient's core beliefs (e.g. that a particular bodily feature is deformed) and ask them several questions about it, such as their degree of certainty that the belief is accurate. The interviewer scores the patient on each question (e.g. from completely convinced that the belief is false to completely convinced that it is accurate). Each item is rated from 0 to 4, from least to most severe, and the total is added up for a composite score.

2004). Researchers also find that referential thinking is typically associated with BDD, but not OCD (Phillips, Pinto, Hart, et al. 2012; Toh, Castle, Mountjoy, et al. 2017). Referential thinking, or the tendency to interpret innocuous, neutral stimuli as having some unique personal meaning, also contributes to one's overall level of delusionality. While referential thinking is a commonly attested clinical feature of BDD, it is not considered typical of OCD (Phillips, Wilhelm, Koran, et al. 2012, 1294).

In addition to a difference when it comes to levels of insight or delusionality, there is also a difference when it comes to the patients' attitudes toward their own behaviors. People with BDD engage in compulsions but their obsessions are not usually as intrusive or resisted as staunchly by themselves as those of individuals with OCD, and those with BDD do not typically regard their beliefs as senseless (Oldham, Hollander, & Skol 1996). This relates to a distinction between "ego-syntonic" beliefs and behaviors, which the individual endorses, and "ego-dystonic" ones, which are not endorsed by the individual himself or herself. While the former are more closely identified with BDD, the latter are typically associated with OCD. There would seem to be a connection between the type and degree of delusion experienced and the syntonicity or dystonicity of the beliefs and behaviors, since those individuals who are unable to acknowledge the senselessness of their own beliefs are likely to be more invested in them and hence to endorse their compulsive behaviors, which is the case with BDD. Meanwhile, individuals with OCD, who tend to have greater insight, are less prone to endorsing their behaviors and associated beliefs.

Another point of contrast between BDD and OCD patients is that the compulsive behavior of BDD patients seems not to relieve their anxiety, unlike the behaviors engaged in by OCD patients. According to Veale and Riley (2001), mirror checking does not result in a reduction of the anxiety experienced by BDD patients. By contrast, the compulsive behaviors associated with OCD, such as repeatedly checking the stove to make sure that it is not on, or washing hands numerous times to ensure that they are not contaminated, seem to relieve the anxiety of OCD patients at least partially and temporarily. Veale and Riley (2001) report that prior to mirror gazing, BDD patients are driven by the hope that they will look different, the desire to know exactly how they look, a belief that they will feel worse if they resist gazing, and the desire to camouflage themselves, but after mirror-gazing, participants feel worse than they did before. Further, Phillips, Wilhelm, Koran, et al. (2010, 578) state that "some BDD compulsions (e.g. mirror checking) do not appear to follow a simple model of anxiety reduction, which is more commonly seen in OCD."

A final contrast between BDD and OCD patients concerns the difference in their respective abilities to understand social situations. As already mentioned, at least in some tasks involving social cognition, BDD patients perform worse than healthy controls, particularly in interpreting the thoughts of others. This difference also holds when they are compared with patients diagnosed with OCD. Those with BDD are found to be less accurate overall in interpreting social situations compared to those with OCD (Buhlmann, Wacker, & Dziobek 2015). This difference appears related to two attributes of people with BDD already mentioned: their level of insight into their own condition and their tendency to engage in referential thinking. BDD patients suffer from delusions about their appearance or harbor entrenched misguided beliefs that they endorse. These beliefs appear to be partly induced, or at least maintained, by deficits when it comes to interpreting the thoughts of others, since they often think that others are judging them negatively when they are not, and they are not easily persuaded by others when they reassure them about their own appearance. Hence, it seems as though deficits in social cognition are central to BDD and are related to some other aspects of the disorder such as the level of delusionality associated with it.

BDD and OCD both involve behaviors that seemingly cannot be stopped (cf. Abramowitz 2018), but they differ when it comes to the individuals' level of insight or delusionality, the degree to which individuals endorse their behaviors and beliefs, the extent to which individuals' behaviors relieve their anxiety, and the individuals' ability to interpret social situations. By emphasizing the internal mental states of BDD patients rather than their external symptoms or behaviors, some important contrasts emerge between BDD and OCD that appear central to understanding BDD. To further justify this claim, we will propose a tentative causal model of BDD in the next section, which further brings out its differences with OCD and suggests reclassifying it.

8.4 Proposal for a Causal Model of BDD

By paying greater attention to the internal mental states of people with BDD rather than their outward symptoms, it becomes apparent that there are important differences between BDD and OCD. We conjecture that these differences point to different causal models of the two disorders. On the surface, it seems as though BDD and OCD patients exhibit similar symptoms of compulsivity and repetitive behaviors, but when one examines the symptoms more closely, it is plausible that these compulsive behaviors involve very different causal processes. The main components of this causal network consist in some of the central characteristics of BDD that set it apart from OCD: focusing on the content of the disorder as opposed to its form; having a lower level of insight; having an ego-syntonic perspective on one's experiences; and the behaviors' not having an anxiety reduction function. These differences can be used to sketch a speculative causal model of BDD, as follows. Two of the most basic cognitive features of BDD patients mentioned in Section 8.2 are: (i) deficits in visual processing; (ii) deficits when it comes to evaluating the emotions, thoughts, and intentions of others. As mentioned earlier, the perceptual deficits involve focusing on details at the expense of a global or configural picture. Both clinical observations as well as neurobiological and psychophysical research suggest that individuals with BDD tend to focus on the details of their appearance at the expense of their overall image, and there is some evidence to suggest that abnormalities in visual perception mechanisms are involved in the onset and maintenance of BDD (see Beilharz, Atkins, Grace, et al. 2016). This implies that individuals are likely to concentrate on specific body parts or features rather than their overall image. Excessive scrutiny of a specific feature may lead in turn to increased dissatisfaction with that feature. Meanwhile, theory of mind deficits involve misinterpreting others' expressions or statements. While these misinterpretations need not be negative or directed at oneself, it is possible that given that people with BDD are dissatisfied with some aspect of their appearance, these theory of mind deficits may lead to a perception of negative evaluation by others. In addition, there may be an interaction between these two posited causal antecedents of BDD, since dissatisfaction with one's appearance may be reinforced by a mistaken interpretation of the attitudes of other individuals. This may cause increased dissatisfaction with one's appearance, which may lead in turn to further negative construals of other people's attitudes. These negative impressions, of one's own appearance and of the evaluations of others are likely to reinforce one another. The positive feedback loop may result in beliefs that are difficult to dislodge because of their self-reinforcing nature, and this is a mark of delusional thinking, as defined by the DSM-5 ("fixed beliefs that are not amenable to change in light of conflicting evidence"). However, in order to give rise to full-blown delusions, an additional causal factor may be required in the form of a disposition toward a certain cognitive bias, as we will try to explain.

Recent research has suggested that there is a causal relationship between a bias against disconfirmatory evidence (BADE) and delusions in a number of psychiatric disorders. As argued in Chapter 7, BADE may be one

8.4 Proposal for a Causal Model of BDD

manifestation of a myside heuristic that is attested in a very wide range of human subjects and it occurs when people reject or discount evidence that is contrary to their beliefs. People who are prone to this cognitive bias tend to discount evidence against their beliefs to a greater extent than may be warranted by ideal standards of rationality. As mentioned in Chapter 7, Woodward, Moritz, Cuttler, et al. (2006) present evidence that psychiatric patients with delusions differ from controls in this regard, exhibiting an accentuated or extreme version of BADE.[8] Moreover, they suggest that BADE is not just a feature of the delusion itself but that it extends to "delusion-neutral material" in experimental tasks. At least in schizophrenic patients, the tendency toward BADE exists independently from the delusion and may play a role in causing and maintaining the delusion. This causal association between BADE and delusions has also been found in other psychiatric disorders. On the basis of a meta-analysis examining the relationship between two closely related cognitive biases, BADE and Jumping to Conclusions (JTC),[9] McLean, Mattiske, and Balzan (2017, 345) conclude: "The association of these biases with delusions in multiple diagnoses would demonstrate this relationship is not limited to schizophrenia, and would support a causal relationship."[10] In addition, BADE has been shown to be associated with delusions in both clinical and nonclinical populations (Bronstein & Cannon 2017). On the basis of these and similar findings, we conjecture that BADE might be involved in the inception and

---

[8] As mentioned in Chapter 7, whether the bias as it is manifested in individuals with delusions is just a heightened form of the bias in controls or whether it is qualitatively different is an open question and one that we will not try to resolve. In what follows, we intend BADE to refer to the version of the bias as it occurs in delusional patients, whether it is just a heightened form of the same bias or an extreme variant.

[9] People who are prone to JTC bias make decisions based on relatively insufficient evidence. The main way to test for this bias involves probabilistic reasoning. Participants are told that they will be shown a sequence of colored beads drawn either from a jar that has 85 percent red and 15 percent black beads, or a jar that has 15 percent red and 85 percent black beads. They are shown one at a time, supposedly randomly but in fact in a predetermined sequence, and told to stop the experimenter when they are confident which jar is being drawn from. Those who are especially prone to JTC make a decision based on seeing significantly fewer beads than the general population.

[10] They also write: "Our results are consistent with the hypothesis that cognitive biases play a causal role in delusions. The hypothesis finds indirect support in treatment studies also. Metacognitive Training, a cognitive therapy that focuses on reducing JTC and BADE, has been shown by meta-analysis to weaken delusional severity in people with schizophrenia" (McLean, Mattiske & Balzan 2017, 352). In an earlier meta-analysis of a number of studies, Fine, Gardner, Craigie, et al. (2007) find that JTC but not BADE is associated with individuals with delusions, but the more recent work just cited seems to undermine their conclusion. They also suggest that the two biases tend to work in opposite directions, since those who are prone to jumping to conclusions may be likely to embrace contrary evidence rather than reject it, but their analysis does not seem to distinguish between attitudes toward one's own beliefs and attitudes toward contrary evidence.

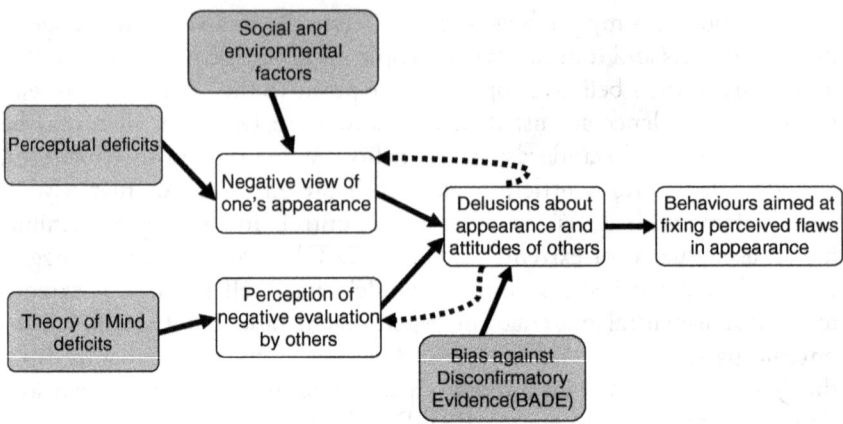

Figure 8.1. Causal model of *body dysmorphic disorder*, showing ultimate causes (grey) and proximal causes and effects (white). Dotted arrows indicate causal feedback loops.

maintenance of delusions present in BDD. In our causal model, we have proposed that perceptual deficits and theory of mind deficits are two of the principal causal factors in BDD, and this additional factor, a disposition toward BADE, may also be causally implicated in causing and maintaining the disorder, specifically in ensuring that the faulty beliefs of BDD patients rise to the level of delusions (see Figure 8.1). People with BDD do not believe others when they try to reassure them that their particular body part is not deformed and they are generally resistant to disconfirming evidence, as we have seen. As already suggested, in some experimental paradigms, most individuals, not just psychiatric patients, are shown to have a bias against disconfirming evidence. However, in individuals with BDD (and in psychiatric patients with delusions more generally), that tendency may exist in a more pronounced and severe form, and when combined with the other causal factors mentioned, may result in the "perfect storm" that is responsible for the emergence of full-fledged BDD. Moreover, since their delusions are rooted in distorted perceptions, BDD delusions are not completely unmoored from reality and hence do not resemble the "bizarre" delusions associated with schizophrenia and other psychiatric disorders. Still, BDD patients maintain certain entrenched beliefs about their appearance that are resistant to change on the basis of contrary evidence.

Now that we have sketched out a causal model of BDD, we will argue that these components render it a psychiatric kind in its own right and make it categorically distinct from OCD. Our causal model of BDD theorizes that it is at least partly caused by a combination of cognitive and perceptual factors. We have suggested that a perceptual deficit, a theory of mind deficit, and an accentuated cognitive bias against disconfirming evidence are crucial causal antecedents of BDD. In addition to these endogenous causal factors, there may also be various social causes that might contribute to the disorder, either directly or indirectly by influencing these psychological causes. The causal model associates BDD with a distinctive network of effects, since the cognitive deficits interact in such a way as to generate delusional beliefs about one's physical appearance. These delusional beliefs, in turn, generate a suite of behaviors that aim at fixing perceived flaws in one's physical appearance, and these behaviors ultimately do not reduce their anxiety. Finally, the beliefs and behaviors involved are such that they constitute a significant departure from rationality and disrupt the functioning of individuals who are characterized by these mental states and behaviors. Departures from rationality, represented primarily by delusional beliefs, lead to disruptive mental states and behaviors, which hamper the ability of individuals with BDD to achieve their goals, avoid suffering, and establish meaningful relationships.[11] This set of characteristic properties constitutes the causal profile of BDD and distinguishes it as a real kind in the psychiatric domain.

One obvious way that BDD differs from OCD is that BDD involves a perceptual/visual deficit, and OCD does not appear to do so (cf. Kaplan, Rossell, Enticott, et al. 2013). In addition, as already seen, persons with BDD differ from those with OCD when it comes to theory of mind deficits. Moreover, there does not seem to be any evidence that OCD patients have a cognitive bias against disconfirmatory evidence. Indeed, there is evidence that people with OCD have the opposite tendency: They consider alternative possibilities more than controls and may be less discriminating in giving them importance (Dèttore & O'Connor 2013; for a review of cognitive biases associated with BDD, see Hezel & McNally 2016). Hence their causal antecedents differ. There are other important ways in which BDD differs from OCD, at least some of which seem to issue from these causal factors. As already mentioned, people with BDD are more focused on their image or particular body part (content), while people with OCD are more concerned with the intrusive preoccupations or thoughts themselves (form) (Rivera & Borda 2001). Someone with BDD may be concerned about the size or shape of their nose, thinking that it is truly misshaped even though

---

[11] A recent literature review finds that BDD patients have high lifetime rates of psychiatric hospitalization (48 percent), suicidal ideation (45 to 82 percent), and suicide attempts (22 to 24 percent) (Mufaddel, Osman, Almugaddam, et al. 2013).

it is not; while someone with OCD is likely to be concerned about the preoccupations of their compulsion, even if they acknowledge that their compulsion is nonsensical. People with BDD and OCD differ not only in their focus, but also in their level of insight. People with OCD tend to have greater levels of insight than those with BDD, which is to say that they are not as delusional. People with OCD can acknowledge that their desire to check the stove multiple times in succession might not make sense, but they still have a desire to do it. People with BDD think that it is perfectly sensible to engage in their "checking" and "fixing" behaviors. When we look at the disorders from the patient's experiential perspective, it seems that OCD is ego-dystonic, while BDD is ego-syntonic.

Since both the experiences and underlying causal factors associated with BDD differ substantially from those of OCD, we posit that there is no reason to group them in the same category of OCRDs. The experiential differences between the disorders might also mean that it does not make sense to use the same treatment methods for them. The causal model of BDD is important for theoretical and classificatory purposes, but it is also possible that examining etiology could lead to implications for treatment. Treatments and therapies that take causes into account might be more effective than those which do not. We will take this issue up in the following section.

8.5 Objections and Replies

In this section, we will consider three objections to our proposed causal model of BDD. The first objection would question the appropriateness of the causal model that we have proposed. Why single out these features in particular as the crucial causal factors for the emergence and maintenance of BDD? Could an alternative causal model adequately explain the features of BDD? The most explicit causal model of BDD that we have found in the psychiatric literature has been proposed by Veale (2004) (see Figure 8.2.). His model involves seven key components: triggers, negative appraisals of an internal body image, safety behaviors, mood, rumination, processing the self as an aesthetic object, and in a separate section, he mentions selective attention. In the graphic illustrating the model, each of these components is connected by bidirectional arrows, and it is difficult to discern the precise nature of the causal relationship between these components. While it is true that the features mentioned are somehow involved in BDD, the model itself is not clear in explaining how they relate to each other. For example, Veale (2004, 114–115) describes the link between a trigger and a negative appraisal as follows: "It is proposed that the cycle

8.5 Objections and Replies

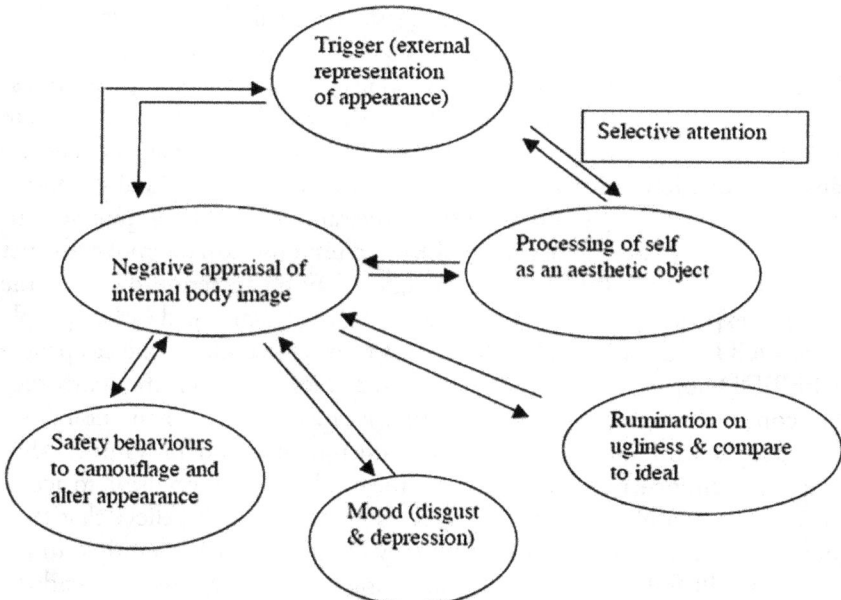

Figure 8.2. A cognitive behavioral model of body dysmorphic disorder proposed by Veale (2004)

begins when an external representation of the person's appearance (e.g. looking in a mirror) activates a distorted mental image." But it is not clear why looking at oneself in the mirror leads to a negative appraisal of one's internal body image and Veale's model does not seem to explain this. By contrast, our proposed model highlights the fact that a perceptual deficit, which involves focusing on details at the expense of a configural picture, may lead to exaggerated attention toward one or more bodily features, resulting in an initial negative self-perception, which is then reinforced by a theory of mind deficit. More recently, Feusner, Neziroglu, Wilhelm, et al. (2010) also try to enumerate some of the main causal factors for BDD, but although they identify the perceptual processing and theory of mind deficits that we have emphasized, in addition to neurochemical and social factors, they do not mention the cognitive bias against disconfirmatory evidence, nor do they attempt to show how these factors might fit together in a more comprehensive causal model.

A second objection questions our understanding of the mental states of BDD patients and sees greater continuity between BDD and OCD. Like us, Abramowitz (2018) suggests that the OCRD category should be reevaluated; however, unlike us, he suggests that of all the OCRDs, BDD is the most similar to OCD. He suggests that OCD compulsions act as a safety behavior, where the behavior allows for one to temporarily escape from distress. He claims that for both OCD and BDD, the patient's thoughts are intrusive and anxiety provoking, and that the repetitive behaviors have an anxiety reduction function. And, the checking behavior maintains appearance-related preoccupations in BDD in the same way that compulsive rituals maintain obsessive fears in OCD. Our response to Abramowitz's first point is that, while the intrusive thoughts may be anxiety provoking, the patients experience their anxiety differently. As we mentioned earlier, people with OCD find the thoughts themselves to be concerning, whereas people with BDD appear not to be concerned that they have such thoughts, they are concerned about the body part that is the subject of their thoughts.[12] For people with OCD, the thoughts are intrusive and unwanted. They might not endorse those thoughts, but they still have a compulsion to act on them. For example, people with OCD may not strongly believe that their house will in fact burn down unless they check their stove multiple times, but they still feel the need to check. People with BDD, however, endorse their thoughts. For them, it is not that the thoughts or preoccupations are unwanted or intrusive, but the "flawed" body part that is unwanted. Our response to Abramowitz's second point that both OCD and BDD involve anxiety reduction functions, is that there is a body of evidence already cited (e.g. Veale & Riley 2001; Phillips, Wilhelm, Koran, et al. 2010) that strongly suggests that this is not the case. As mentioned earlier, people with BDD do not feel as though they have completed their goal after engaging in their BDD behaviors in the same way that people with OCD have completed their goal once their ritual has been performed and their anxiety has been reduced. To Abramowitz's third point that the checking behavior maintains appearance-related preoccupations in BDD in the same way that compulsive rituals maintain obsessive fears in OCD, we respond by saying: *That* the behavior is maintained is different from *why* the behavior is maintained. We do not deny that both people with BDD and people with OCD engage in checking behaviors, and deal with impulse-control issues. What we have argued is that these issues are experienced differently, and that these behaviors have a different etiology.

Finally, it might be objected that a theoretical causal model should not be the basis for classifying psychiatric disorders, including BDD. Indeed,

---

[12] One literature review found that between 6 and 15 percent of patients seeking cosmetic surgery have BDD (Mufaddel, Osman, Almugaddam, et al. 2013).

it might be said that the basis for classification in clinical psychiatry should be treatment[13] not causal modelling, and that classification ought to be aimed at remedying disorders rather than explaining them. But we doubt that there is such a stark contrast between the aims of treatment and causal explanation. The causal model that we have proposed attempts to identify some of the causal factors that contribute to the emergence of the disorder in people with BDD. Some treatments may be found to be efficacious without understanding the reasons for their efficacy, but these will remain lucky guesses unless the causal process that underlies them is understood. We think that it is unlikely that an informed treatment of psychiatric disorders can refrain from intervening on their causal antecedents in order to attempt to avert the causal process that leads to the emergence of the disorder. While knowledge of the causal process may not always lead to effective treatments, perhaps because it is impracticable to intervene on the relevant causes, such an intervention would seem necessary to an informed treatment (as opposed to a mere guess).

8.6 Conclusion

There is growing interest in psychiatric research and in philosophy of psychiatry in understanding the etiology of mental disorders in terms of specific cognitive dysfunctions or deficits.[14] In this chapter, we have tried to synthesize some of the recent research on Body Dysmorphic Disorder to emerge with what we consider to be a plausible causal model of the disorder. The primary causal factors that we posit as part of the etiology of BDD, namely a deficit in visual perception, a theory of mind deficit, and a form of the cognitive bias against disconfirmatory evidence, are certainly not the whole story. But if something like this causal network lies behind BDD and can be used to explain its emergence and persistence, that would support the case for considering it a real kind. Though not exclusively cognitive, this kind has an important cognitive dimension. In previous chapters, the case has been made that many cognitive kinds are likely to resist reduction to neural, neurochemical, or neurophysiological constructs, mainly because they are individuated contextually, etiologically, or both. In this case, one obstacle to a reductive account is the occurrence of a cognitive bias in the causal network. If, as suggested in Chapter 6,

---

[13] Or more cynically, in the service of practical purposes like insurance payments (see e.g. Greenberg 2013).
[14] See particularly the work of Garety and collaborators, for example, Garety, Bebbington, Fowler, et al. (2007).

the identification of a cognitive heuristic as a bias may require relating it to a broader environmental and social context, then BADE cannot be characterized as such in isolation. In addition, since BADE is thought to be present in an extreme form in individuals with BDD, the severity of this bias is also a contextual matter and cannot be identified without such a context. Moreover, it would seem as though identifying BDD as a *disorder* is dependent on discerning its effect on the behavior of the individual in a social context. That is, individuals who have this particular combination of psychological traits might not be singled out as a particular human type if it were not for the harms that they endure and the difficulties they face in achieving their goals in a social setting. Still, it might be objected, the causal network that we have identified would seem to constitute a kind in its own right whether or not it has harmful effects on the individuals who possess them. There is something to be said for this view, particularly since we raised doubts in Section 8.1 that the superordinate category of *psychiatric disorder* might correspond to a kind. Therefore, we would not rule out the possibility that the components of the causal network that we have identified with BDD may each be identified with neural correlates (though they might be multiply realizable neurally), and that their combination would also correspond to a set of neural correlates. Nevertheless, it is the combination of these causal factors that issues in BDD and there is no reason to think that this collection of neural correlates has any claim to unity beyond their issuing in a set of cognitive and behavioral effects. This means that the psychological causal network provides the basis for singling out this collection of features and considering it a unified kind.

Finally, what kind of kind is BDD: Is it a state, event, process, capacity, or what exactly? Though there are events and processes involved in the etiology of BDD as we have described it, as well as (dysfunctional) capacities, the most natural understanding of the category is that it identifies a kind of *individual*: a person with BDD. The causal process that we have identified takes place in an individual and results in a kind of person who is prone to certain behaviors (e.g. mirror checking) and to having certain states of mind (e.g. delusions about appearance), and hence difficulties in achieving their goals and flourishing in a social setting. It is this kind of person that is primarily a member of the kind *BDD*, though to say that BDD characterizes a type of person is not to say that persons with BDD are essentially or irreversibly persons of that kind.

CHAPTER 9

Epilogue

... the great regions of the mind correspond to the great regions of the brain.
> – Paul Broca, "Remarks on the Seat of the Faculty of Articulate Language, Following an Observation of Aphemia (Loss of Speech)"

... when you are a Bear of Very Little Brain, and you Think of Things, you find sometimes that a Thing which seemed very Thingish inside you is quite different when it gets out into the open and has other people looking at it.
> – A. A. Milne, *The House at Pooh Corner*

9.1 Introduction

The aim of this book has been to provide an account of cognitive ontology, but rather than try to give some kind of comprehensive tour of the real kinds of cognition – if that were even possible – my approach has been to select a limited number of (what I take to be) both significant and representative taxonomic categories and use them to illustrate and support a theory of cognitive kinds, or real kinds in the cognitive domain. The general theoretical framework of real kinds was sketched out in Chapter 1 and it was applied to a variety of case studies in successive chapters of this book. Some of these cases are household categories that will be familiar to laypersons and cognitive scientists alike, such as *concepts*, *innateness*, and *(episodic) memory*, but others are not so well-known, for example, *language-thought processes* and *cognitive heuristics*. This is to be expected, since the aim is to choose some categories that are good candidates for kinds as well as some that are not, and the latter are less likely to be widely known or recognized. The fact that I had to come up with a label for "language-thought processes" is itself an indication that it is not usually treated as a cognitive kind in its own right. Still, analyzing this category is not just an exercise in tilting at windmills, since the ways in which it fails to be a cognitive kind are instructive and illustrate certain general principles for identifying kinds in the cognitive sciences. The same goes for the category of *cognitive heuristics*. Moreover, in these cases and some others, even when a particular category is not a good candidate for being a cognitive kind, there are others in the vicinity that are (e.g. *myside heuristic*).

There is a danger in a book like this, containing successive chapters tackling separate case studies, that no single argumentative thread will emerge and that there may not be enough continuity among the cases to unify the whole. I have already attempted to indicate links and parallels in previous chapters, when they arose, but in this brief epilogue, I will try to draw further connections between chapters, and highlight certain common themes that bear emphasizing. In particular, in what follows, I want to draw attention to three main lessons that arise from these case studies. The first concerns the etiological–environmental individuation of many cognitive categories (Section 9.2), the second has to do with broader ontological issues in cognitive science (Section 9.3), and the third has to do with reductionism and the status of cognitive neuroscience (Section 9.4).

9.2 Etiological–Environmental Individuation

The categories discussed in this book pertain specifically to human cognition, not to psychology broadly understood. There are commonalities among categories in the cognitive domain that distinguish them from, say, categories in the domains of perception and emotion. These commonalities are engendered by the fact that cognition seems to delineate a relatively "closed system" (in the sense of Chapter 1). There are surely dense connections between cognition on the one hand and perception and emotion on the other, but the division between them is not an artificial one.[1] The categories discussed in previous chapters all pertain to the cognitive domain, though some of them are also implicated in processes involving perception and emotion, not to mention processes that lie outside the psychological domain altogether. For example, perceptual states serve as inputs to the capacity of *episodic memory*, and there is some evidence that the *myside heuristic* is influenced by affective processes. There is also evidence that *innate cognitive capacities* have a genetic basis. But these cognitive kinds

---

[1] See Beck (2018) and Phillips (2019) for two recent philosophical proposals for drawing the perception-cognition boundary on principled grounds, involving (respectively) stimulus-independence and stimulus control.

9.2 Etiological–Environmental Individuation

can be investigated using proprietary methods distinctive of cognitive science, and the causal processes in which they feature can be demarcated from others for the purposes of many inquiries.

One aspect of several of these cognitive kinds that I have tried to emphasize throughout is their externalist aspect, or as I have put it, their etiological–environmental individuation. This cumbersome expression is a more accurate way of describing the taxonomic practices that I have in mind for reasons that bear repeating here. It is misleading to describe the factors with reference to which some cognitive kinds are individuated as "external" primarily because the term "external" is insufficiently descriptive and does not indicate the precise features of reality that serve to ground these cognitive categories. In many of the cases discussed, what matters most is etiology: the causal history of the entities in question. Moreover, the causal history in question sometimes pertains to the developmental history of the individual and at other times the history of the species or phylogenetic lineage. The "externalist" label is not an apt epithet for these features of reality, but I will use it here because it has become so entrenched.

Externalism is commonly regarded as an established thesis in the philosophy of mind (though it is not without its detractors), but I will try in this section to distinguish the view defended here from other varieties of externalism in the philosophy of mind and cognitive science. As is widely known, there are two major philosophical proposals that emphasize the role of the "external world" in understanding the mind. The first is the externalist thesis associated with the work of Putnam, which is colloquially encapsulated in the expression, meanings "just ain't in the head" (1973, 704). The second is the "active externalism" or "extended mind thesis" associated with the work of Clark and Chalmers (1998), which holds that mental states and processes are characterized by "reliable coupling" with features of the environment. Starting with the second thesis first, the environmental–etiological thesis that I have advocated is not just about causal coupling but about the very identification or individuation of cognitive kinds. I have tried to show that many features of cognition are individuated with reference to their environment or etiology. This claim is more far-reaching than "active externalism" since it says that when it comes to something like a state of episodic memory (say), it is not even possible to identify it as such without invoking its causal history. If causal factors in the past or current environment of the thinker determine whether a certain cognitive state is a memory state in the first place, then it follows that memory states do not supervene on the thinker narrowly conceived. Hence, this thesis is stronger than one that claims merely that cognitive states of the thinker are causally coupled with such entities as notebooks and smartphones (Clark & Chalmers 1998; Chalmers 2008). Indeed, it means that at least for some cognitive kinds, it is not clear how to frame the thesis that "internal" states are coupled with "external" entities, since it is misguided to talk about cognitive states being internal in the first place. As for the second familiar externalist thesis, it is explicitly an individuative thesis, but it is usually restricted to meanings, concepts, or semantic content. Concurring with this claim, I have argued in Chapter 2 that externalism holds for *concepts* (though I think they are individuated on the basis of both externalistic and internalistic factors), but I have also put forward reasons for thinking that it applies to other cognitive kinds, such as *innateness*, *domain specificity*, *episodic memory*, and *myside bias*. Moreover, the arguments I have made rely not on intuitive judgments concerning the ascription of conceptual contents, but taxonomic practices in cognitive science. For example, when it comes to concepts, I have tried to present empirical evidence showing that much of the empirical work in developmental psychology individuates the concepts of concept learners (partly) according to their etiology in the natural and social worlds (see Section 2.4). Though his emphasis has been on perception, Burge (1986; 2010) has also argued that externalism applies to kinds in scientific psychology. Based on this claim, he has made the most explicit case for the conclusion that the taxonomic kinds of psychology may not coincide with the kinds of physiology:

> Without a set of physical transactions, none of the intentional transactions would transpire. But it does not follow that the kinds invoked in explaining causal interactions among intentional states (or between physical states and intentional states – for example, in vision or in action) supervene on the underlying physiological transactions. The same physical transactions in a given person may in principle mediate, or underly [sic], transactions involving different intentional states – if the environmental features that enter into the individuation of the intentional states and that are critical in the explanatory generalizations that invoke those states vary in appropriate ways. (Burge 1986, 17)

My claim is that the etiological–environmental individuation of cognitive kinds is not only directly supported by taxonomic practices in cognitive science, but is also more far-reaching than many externalists have tended to assume. Moreover, some of the implications of externalism do not appear to have been widely acknowledged. In particular, apart from occasional stray remarks like Burge's just cited, it is not generally appreciated that the environmental–etiological individuation of cognitive categories may scuttle neat matchups with neural categories.

9.2 Etiological–Environmental Individuation

So far, I have emphasized etiology in the ontogenetic sense, but phylogenetic etiology is also relevant to cognitive science. For instance, I argued in Chapter 5 that what makes something a system or capacity of episodic memory is at least partly its function or what it evolved for. This means that in deciding whether nonhuman animals have episodic memory, we would be guided at least in part by the etiology of the systems or capacities.[2] More blatantly, when determining whether a cognitive capacity is domain-specific or not, I argued that this can only be done against a background of selection history. According to the account put forward in Chapter 4, the domain specificity of a cognitive capacity can only be identified with reference to the function that it evolved to perform. At least, that seems to be the most defensible understanding of the cognitive kind, as it is ordinarily deployed in cognitive science. These cognitive kinds are individuated primarily by their phylogenetic etiology, though that is not the only factor used to identify them. For example, domain-specific cognitive capacities are ones that do not generalize beyond their proper function, which means that they are individuated with reference to their synchronic causal powers as well as their etiology.

Etiological individuation is not the only type of relational individuation in cognitive science; in some cases, the synchronic environment of the thinker also plays a role in determining the identity of the cognitive kind that is manifested. For example, the central property in the causal cluster comprising the kind *innate cognitive capacity* is *triggerability*, which makes implicit reference to the informational content of the stimulus in the thinker's environment. Additionally, in the case of the *myside heuristic*, it was argued in Section 7.4, that determining whether an instance of the heuristic is indeed a *bias* requires doing so against the background of a particular cognitive task or problem. Hence, if researchers want to distinguish a myside bias from a myside heuristic, they can only do so with reference to the synchronic context of the thinker.

In the case of *concepts* and in some other cases, I have tried to argue that the kinds involved are individuated in part according to their etiology and in part according to their causal powers. They can be thought of as hybrid kinds. Generally speaking, if an entity has a certain causal history, that does not thereby confer on it any synchronic causal powers. For instance, there is nothing in common to all the human babies born on

---

[2] That need not mean that the only episodic memory capacities are homologues; they may also be analogues, where the functions are individuated at least partly etiologically. See Section 4.5 for further discussion.

January 1, 2000 (that they do not share with babies born on other days). Yet, in many cases in the biological domain, entities that share a history do share many synchronic properties because they have been produced by a causal process or mechanism whose function it is to generate entities with certain properties. A subset of such processes can be characterized as "copying processes" (Millikan 1999), although they need not copy entities perfectly – nor might that be their function. The outputs of such processes are what she calls copied kinds, including biological species and industrial artifacts. This phenomenon seems especially significant in the realm of cognition, where etiology often goes hand in hand with informational or representational content. When it comes to concepts in particular, a great deal of energy has been expended by philosophers on cases in which causal history pulls in one direction while causal powers push in another. These are cases in which the content of a thinker's concept has the "right" etiology but "wrong" causal powers, or vice versa (e.g. the famous Twin Earth cases). What ultimately determines concept identity in such cases? I indicated in Chapter 2 that there may not always be a determinate answer to such questions. One way to think about these cases is to regard them as inevitable outcomes of the convoluted causal structure of the world, which includes causal patterns that combine synchronic and diachronic elements. In discerning these patterns, there are sometimes various ways in which taxonomic lines can be drawn while preserving causal structure. That is why, in one sense, and for some purposes, the Earthian and Twin Earthian share a concept, while in another sense and for other purposes, they do not. But rather than bifurcating concepts to account for bizarre cases, it seems more parsimonious for cognitive scientists not to multiply concept categories beyond necessity, and deploy a hybrid category that is sensitive to both causal history and causal power. I have argued that that is what many cognitive scientists investigating concepts tend to do, albeit implicitly. This is notably the case in developmental psychology where concept possession, concept sharing, and conceptual change are of paramount importance.

This approach to cognitive taxonomy is in keeping with the characterization of the inaptly named "computational level," proposed by Marr (1982). As understood in this book, and as argued in previous chapters (see especially Section 1.5), I take the study of cognition to be conducted largely within Marr's computational level of analysis or explanation. By contrast with an algorithmic or implementational theory, a computational theory asks: "What is the goal of the computation, why is it appropriate, and what is the logic of the strategy by which it can be carried out?" (Marr

1982, 25) Thus, when it comes to vision, Marr (1982, 23) characterizes a computational theory as follows: "In the theory of visual processes, the underlying task is to reliably derive properties of the world from images of it." The corresponding task for the theory of conceptual processes might be characterized as: to reliably generalize from particulars in the world so as to enable recognition, categorization, inference, and action. These accounts are broadly functional in character, where function is understood both synchronically and diachronically, and where functions allude to the worldly task or problem that the cognizer is solving or has evolved to solve.

9.3 Ontological Categories

In previous chapters, I have conjectured that the overarching or higher-order ontological categories characteristic of cognitive science are: *individuals* (or *objects*), *states*, *processes*, and *capacities*. At least, these are the broad ontological categories that include the kinds that have been explicitly discussed in this book. That is not to say that we might not need to posit other higher-order ontological categories, such as events or dispositions, in cognitive science. But it does seem as though these four categories capture almost all the kinds discussed in previous chapters. In this list of broader ontological categories, I have not included the category *mechanism*, which is conspicuous mainly by its absence in previous chapters. Many philosophers would expect mechanisms to play a leading role in cognitive ontology, but I would argue that they are more prevalent in the neural sciences than in the cognitive ones proper. Some philosophers might object that this unnecessarily narrows the scope of the *mechanism* category and its associated explanations, but what seems valuable about the category of *mechanism* is that it identifies a set of distinctive phenomena that lie within the causal sphere. When "mechanism" is understood in such a way as to encompass causality in general, we appear to lose this added descriptive and explanatory content.[3] The type of ontological configuration identified by many philosophers who have written about mechanisms and mechanistic explanation paradigmatically involves structures with proper spatial

---

[3] For an authoritative treatment of mechanisms, but one that in my view over-extends the concept, see Glennan (2017). For example, he states: "What is the relationship between mechanisms and causation? Put briefly, it is just that causes and effects must be connected by mechanisms" (Glennan 2017, 145). For a related criticism of some recent work on mechanism, see Ross (2020, 143), who writes: "A philosophical account that collapses these distinctions [e.g. between causal mechanism and causal pathway], and uses 'mechanism' as an umbrella term for all causal concepts, fails to capture such distinctions and their role in describing and explaining biological phenomena."

parts or components, arranged contiguously, and engaged in activities that involve physical contact. This type of configuration is more characteristic of neuroscience than it is of psychology, whose kinds and explanations tend to be functional rather than mechanistic. Being pitched mainly at the implementational level, many neural kinds are often structurally defined, have relatively well-defined physical boundaries, contain proper physical parts, and interact largely by means of physical contact.

The identification of the higher-order ontological categories that comprise the cognitive domain is one payoff of adopting the perspective of real kinds for the study of cognitive ontology. Another advantage of a real-kind approach is that it can help expose ambiguities in some of our taxonomic categories, for example, *episodic memory*, which can refer to a kind of capacity but also a kind of state. This is not exactly news, since memory researchers have pointed this out in various contexts, but in the case of episodic memory, reflection on this distinction brings out the fact that the capacity is primary, since *states* of episodic memory are such as a result of being produced by the *capacity* of episodic memory. This allows us to avoid tricky questions about the precise synchronic properties that ought to be possessed by states of episodic memory, for example, how much inaccuracy to tolerate in an episodic memory state before ruling it not to be a memory at all. On this approach, an episodic memory state is simply one that is produced by the capacity of episodic memory. Another advantage of a real-kind approach is that by distinguishing superordinate and subordinate categories and kinds, it becomes clearer that, for example, *episodic memory* might be a kind but not the superordinate category *memory*, or that *myside heuristic* might be a kind but not the superordinate category *heuristic*. Moreover, this implies that category labels like "episodic memory" and "myside heuristic" should not be parsed as identifying genus and species, or genus and differentia. The perspective of real or cognitive kinds also makes clearer the relationship between the cognitive kind *concept* and the cognitive kinds corresponding to specific concepts (e.g. APPLE, ORANGE), which I argued, is similar to the relationship between the kind *species* and specific species kinds (e.g. *Panthera tigris*, *Drosophila melanogaster*), rather than the relationship between superordinate and subordinate kinds. In fact, it seems closer to the determinable-determinate relationship (like the relationship of *color* to *red*, or *red* to *scarlet*). This was used to justify thinking of specific concepts as cognitive kinds in their own right.

In previous chapters, *episodic memory* has been classified as a kind of capacity, and so have *innateness* and *domain specificity*. But it might seem as though they are kinds of capacity in quite different senses. If a cognitive capacity has been identified as one of episodic memory that seems to be different from classifying a kind of capacity as innate. Some might say that the former is a true kind of capacity, in the sense that all such capacities, in humans, other animals, and alien life forms, would belong to a single kind, whereas the latter is just a *property* of cognitive capacities. But the alleged contrast is overblown. Compare the following claim: In biological taxonomy, if an organism is classified as a *tiger*, that may seem fundamentally different from classifying it as a *predator* or *carnivore*. But that judgment does not seem warranted, if these are all real biological kinds. A real kind, I have argued, corresponds roughly to multiple (causally connected) properties. Thus, innate cognitive capacities typically share a host of causally connected properties, as argued in Chapter 3, just as episodic memory capacities do. Some kinds of cognitive capacity may be more inclusive than others, but that is no reason not to judge them to be kinds of cognitive capacity nevertheless.

9.4 Reductionism and Cognitive Neuroscience

The idea that cognitive phenomena will correspond in a simple and direct way to neural ones has been aptly labeled the "Simple Coordination Thesis" by Sterelny (2003). According to Sterelny (2003, 7), that thesis says that "meaning is a specific connection property of the wiring-and-connection facts," and he advises us to take seriously the possibility that "the relationship between the two sets of facts is much less clean" than the thesis supposes.[4] This is not to say that neural entities do not have representational properties, just that their representational properties need not translate directly to those of the full-blown cognitive or psychological entities that are associated with them.[5] Throughout this book, in examining taxonomic categories from the cognitive sciences, I have been guided by Marr's distinction between levels of analysis, and have maintained that the computational level is the proper domain of the cognitive. But

---

[4] At the risk of caricature, one can compare the Simple Coordination Thesis to the tendency of seventeenth- and eighteenth-century chemists, who subscribed to the atomic theory, to project from macro-properties of substances to their micro-constituents. For example, Whewell criticizes the French chemist Nicolas Lemery, who subscribed to the idea that acids taste sour because the micro-particles that constitute them have sharp edges, quoting him as writing: "I hope no one will dispute, seeing every one's experience does demonstrate it: he needs but taste an acid to be satisfied of it, for it pricks the tongue like anything keen and finely cut" (cited in Whewell 1840/1847, 382). The claim struck Whewell as both "gratuitous" and "useless."

[5] See Shea (2018) for a recent account of the representational properties of neural entities, which does not, it appears, presuppose something like the Coordination Thesis.

this might not square with a common conception of cognitive science, according to which it comprises such disciplines as neuroscience, or at least subdisciplines such as cognitive neuroscience. In fact, by thinking of the cognitive as a relatively "closed system," I might be accused of denying the very possibility of cognitive neuroscience. That would be an alarmist conclusion for two main reasons. The first is that even though I have argued that there are principled obstacles to reducing many cognitive kinds to neural kinds, that is not to deny that there are at least some reductions in the offing. The argument made in this book against blanket reductionism does not preclude the internalist individuation of *some* cognitive kinds and their identification with neural kinds. Second, and more importantly, just because there are not likely to be many reductions among psychological and neural categories, that is not to say that there are no important links between the respective sciences, indeed even among their categories. As many philosophers of science have noted, scientific disciplines and subdisciplines are related in intricate ways, notably by means of "interfield theories" that do not entail a direct or indirect reduction between them (Darden & Maull 1977). This results in a rather messier picture of the relationship between psychology and neuroscience than tends to be assumed in cognitive neuroscience and related subfields (cf. Stinson 2016). Indeed, it may have a salutary effect on cognitive neuroscience not to orient around the search for *neural correlates* of cognitive constructs. So far, that approach has not reaped dividends and it narrows the scope of expected relationships between the various research programs that investigate the mind–brain.

If neuroscience should not expect to find neural correlates for cognitive categories, how are we to proceed to build bridges between cognitive science and neuroscience? Specifically, what happens to cognitive neuroscience if investigators are not supposed to engage, for example, in reverse inference, which presumes that a particular cognitive capacity is being deployed on the grounds that its associated brain region is active? If there are no neat matchups between cognitive phenomena and neural ones, such inferences would appear to be ruled out and this methodological strategy would need to be revised. But, as I have tried to point out, such matchups are by no means always out of reach. It is just that many of the cognitive kinds investigated here cannot be expected to correlate with neural kinds, and they do not do so for principled reasons that seem to hold for a wide range of kinds in the cognitive domain. Furthermore, this outcome is to be expected given the precedent of other closely related pairs of sciences. Despite the intricate and intimate relationships between ecology and

9.4 Reductionism and Cognitive Neuroscience

genetics, to take a similar pair, there is no expectation that their considered categories will enter into a one-to-one correspondence. To be sure, there is a subdiscipline of ecological genetics, but it does not revolve around the search for genetic correlates of ecological constructs.

It may be objected here that the reason that the cognitive categories encountered in this book may not seem to correlate neatly with neural categories is that they are too mired in a prescientific and unsystematic way of thinking about cognition, including some of the relatively novel categories discussed in this book. Francken and Slors (2014; 2018) make a distinction between "commonsense cognitive concepts (CCCs)" and "scientific cognitive concepts (SCCs)," pointing out that there can be a lack of congruence between them. If many of the categories that cognitive science currently deploys are either identical with commonsense ones, or closely associated with them, we might do well to adopt a wait-and-see attitude before according them their own proprietary domain and declaring their relative autonomy from the implementational level.[6] Moreover, if the categories of cognition do not appear to be converging with those of neuroscience proper, that may be an indication that the former should be abandoned rather than retained.

I would provide two responses to this concern. The first is that even though many of our current cognitive categories may be revised, some of their distinctive features are likely to survive. The environmental–etiological individuation of cognitive kinds has been justified in large part by a certain approach to cognition pioneered by Marr. From this perspective, or so I have argued, the domain of the cognitive incorporates the developmental history of the organism and its phylogenetic lineage, as well as its current environment, in the investigation of cognizant behavior, and indeed in the very individuation of cognitive kinds. This aspect of the methodology and theoretical underpinnings of the study of cognition, is likely to survive the specific categories that cognitive science has devised. The second response speaks more directly to the indispensability of the cognitive domain. It is possible that many of our cognitive categories, particularly those that

---

[6] Indeed, it may be added that we have already begun to speak at least partly in neural terms and to replace some psychological terms with neural ones. However, many if not most such uses in ordinary parlance are clearly unwarranted. Francken and Slors (2018, 70) present interesting examples of misguided cases of "folk neuroscience" (e.g. "the surge in my endorphins was so swift and high that ... I would lose all control"). In my view, there are at least a couple of things wrong with such locutions. First, those making such claims usually have no way of knowing that they are true (e.g. that endorphins actually surged). Second, they often mix and match neural and psychological terms without warrant, since the relationship between neurophysiological and mental phenomena is barely understood at this point.

originate in our folk theories (e.g. *concept*, *innateness*, and *memory*), may need to be discarded to keep up with the inexorable march of science. This may also apply to categories of more recent vintage, which may have been introduced with insufficient warrant (e.g. *domain specificity*, *myside heuristic*, and *Body Dysmorphic Disorder*). Ontological revisionism is a live option when it comes to cognition, as in other domains. Having said that, it cannot be denied that the relationship between our everyday taxonomic categories and those of a science of the mind are likely to be more intimate than with other sciences. In other domains, we can jettison taxonomic categories with abandon. But that is not the case when it comes to psychology or cognitive science. Much of our interest in developing a science of the mind lies in the ability to provide insights into the nature of our mental life. This means that if the categories developed by a mature cognitive science are entirely disjoint with our folk categories, this would not only weaken our interest in developing such a science but it would seem to preclude our using it to help us explain and make sense of our thoughts and actions as we ordinarily conceive of them. This is a common theme in the debate surrounding eliminativism about mental categories: that in eliminating our ordinary mental categories entirely, we would lose the ability to better understand ourselves (see e.g. Baker 2011). It is not inconceivable that we might move beyond the categories that we currently use to explain and predict people's actions and utterances, and that we would largely replace them with a new cognitive lexicon. But the prospects of doing so, at least in a thoroughgoing fashion, seem quite remote. For one thing, as I already indicated elsewhere (see Section 1.5; see also Section 2.4), wholesale conceptual revision is much less common in intellectual history and the history of science than is often supposed. For another, theoretical advances are often made by introducing new categories alongside current ones, or by dividing our current categories into sub categories while preserving the original categories. In other words, a complete overhaul of our cognitive categories is not in the offing, though some conceptual revision in light of empirical inquiry is certainly not to be ruled out.

References

Abramowitz, J. (2018). Presidential address: Are the obsessive-compulsive related disorders related to obsessive-compulsive disorder? A critical look at DSM-5's new category. *Behavior Therapy, 49*(1), 1–11.

Addis, D. R. (2018). Are episodic memories special? On the sameness of remembered and imagined event simulation. *Journal of the Royal Society of New Zealand, 48*(2–3), 64–88.

Aggleton, J. P., & Brown, M. W. (2006). Interleaving brain systems for episodic and recognition memory. *Trends in Cognitive Sciences, 10*(10), 455–463.

Aizawa, K. (2017). Multiple realization, autonomy, and integration. In D. M. Kaplan (ed.), *Explanation and integration in mind and brain science* (pp. 215–235). Oxford University Press.

Aizawa, K., & Gillett, C. (2009). Levels, individual variation and massive multiple realization in neurobiology. In J. Bickle (ed.), *The Oxford handbook of philosophy and neuroscience* (pp. 539–582). Oxford University Press.

American Psychiatric Association. (2013). *Diagnostic and statistical manual of mental disorders (DSM-5®)*. American Psychiatric Press.

Amundson, R., & Lauder, G. V. (1994). Function without purpose: The uses of causal role function in evolutionary biology. *Biology and Philosophy, 9*(4), 443–469.

Anderson, M. L. (2010). Neural reuse: A fundamental organizational principle of the brain. *Behavioral and Brain Sciences, 33*(4), 245–266.

Anderson, M. L. (2014). *After phrenology: Neural reuse and the interactive brain*. MIT Press.

Anderson, M. L. (2015). Mining the brain for a new taxonomy of the mind. *Philosophy Compass, 10*(1), 68–77.

Andreasen, N. C. (1997). Linking mind and brain in the study of mental illnesses: a project for a scientific psychopathology. *Science, 275*(5306), 1586–1593.

Antony, L. (2016). Bias: Friend or foe? Reflections on Saulish skepticism. In M. Brownstein and J. Saul (eds.), *Implicit bias and philosophy, volume 1: Metaphysics and epistemology* (pp. 157–191). Oxford University Press.

Ariew, A. (1999). Innateness is canalization: In defense of a developmental account of innateness. In V. G. Hardcastle (ed.), *Where biology meets psychology: Philosophical essays* (pp. 117–138). MIT Press.

Ariew, A. (2007). Innateness. In M. Matthen and C. Stephens (eds.), *Philosophy of biology (Handbook of the philosophy of science vol. 3)* (pp. 567–584). Elsevier.
Armstrong, S. L., Gleitman, L. R., & Gleitman, H. (1983). What some concepts might not be. *Cognition*, 13(3), 263–308.
Ásta (published under Sveinsdóttir, Á. K.). (2013). The social construction of human kinds. *Hypatia*, 28(4), 716–732.
Athanasopoulos, P., & Casaponsa, A. (2020). The Whorfian brain: Neuroscientific approaches to linguistic relativity. *Cognitive Neuropsychology*, 37(5–6), 393–412.
Atkinson, A. P., & Wheeler, M. (2004). The grain of domains: The evolutionary-psychological case against domain-general cognition. *Mind & Language*, 19(2), 147–176.
Baddeley, A. (2001). The concept of episodic memory. *Philosophical Transactions of the Royal Society of London, Series B*, 356(1413), 1345–1350.
Baker, L. R. (2011). Does naturalism rest on a mistake? *American Philosophical Quarterly*, 48(2), 161–173.
Baker, M. C. (2001). *The atoms of language: The mind's hidden rules of grammar.* Basic Books.
Baldock, E., Anson, M., & Veale, D. (2012). The stopping criteria for mirror-gazing in body dysmorphic disorder. *British Journal of Clinical Psychology*, 51(3), 323–344.
Baluška, F., & Levin, M. (2016). On having no head: Cognition throughout biological systems. *Frontiers in Psychology*, 7, 902.
Barclay, J. R., Bransford, J. D., Franks, J. J., McCarrell, N. S., & Nitsch, K. E. (1974). Comprehension and semantic flexibility. *Journal of Verbal Learning and Verbal Behavior*, 13(4), 471–481.
Baron, J. (2004). Normative models of judgment and decision making. In D. Koehler and N. Harvey (eds.), *Blackwell handbook of judgment and decision making* (pp. 19–36). Blackwell.
Barrett, H. C. (2018). Modularity and domain specificity. In H. Callan (ed.), *The international encyclopedia of anthropology* (pp. 1–10). Wiley.
Barsalou, L. W. (1982). Context-independent and context-dependent information in concepts. *Memory & Cognition*, 10(1), 82–93.
Barsalou, L. W. (1983). Ad hoc categories. *Memory & Cognition*, 11(3), 211–227.
Barsalou, L. W. (1993). Flexibility, structure, and linguistic vagary in concepts: Manifestations of a compositional system of perceptual symbols. In A. C. Collins, S. E. Gathercole, and M. A. Conway (eds.), *Theories of memory*, (pp. 29–101). Lawrence Erlbaum Associates.
Barsalou, L. W. (2016). On staying grounded and avoiding quixotic dead ends. *Psychonomic Bulletin & Review*, 23(4), 1122–1142.
Barsalou, L. W., Dutriaux, L., & Scheepers, C. (2018). Moving beyond the distinction between concrete and abstract concepts. *Philosophical Transactions of the Royal Society B.* 373(1752), 20170144.
Barsalou, L. W., Simmons, W. K., Barbey, A. K., & Wilson, C. D. (2003). Grounding conceptual knowledge in modality-specific systems. *Trends in Cognitive Sciences*, 7(2), 84–91.

Bascandziev, I., Tardiff, N., Zaitchik, D., & Carey, S. (2018). The role of domain-general cognitive resources in children's construction of a vitalist theory of biology. *Cognitive Psychology, 104*, 1–28.

Bates, E. (2001). Modularity, domain specificity and the development of language. In W. Bechtel et al. (eds.), *Philosophy and the neurosciences: A reader* (pp. 134–151). Blackwell.

Bateson, P. & Mameli, M. (2007). The innate and the acquired: Useful clusters or a residual distinction from folk biology? *Developmental Psychobiology, 49*(8), 818–831.

Beck, J. (2018). Marking the perception–cognition boundary: The criterion of stimulus-dependence. *Australasian Journal of Philosophy, 96*(2), 319–334.

Bédécarrats, A. & Glanzman, D. L. (2018). Regulation of memory storage through epigenetic alterations: A new role for RNA. *The Biochemist, 40*(5), 12–15.

Beilharz, F. L., Atkins, K. J., Duncum, A. J., & Mundy, M. E. (2016). Altering visual perception abnormalities: A marker for body image concern. *PloS ONE, 11*(3), Article e0151933.

Beilharz, F., Castle, D. J., Grace, S., & Rossell, S. L. (2017). A systematic review of visual processing and associated treatments in body dysmorphic disorder. *Acta Psychiatrica Scandinavica, 136*(1), 16–36.

Binder J. R. (2016). In defense of abstract conceptual representations. *Psychonomic Bulletin & Review, 23*(4), 1096–1108.

Blanco F. (2017). Cognitive bias. In J. Vonk and T. Shackelford (eds.), *Encyclopedia of animal cognition and behavior*. Springer.

Block, N. (1986). Advertisement for a semantics for psychology. *Midwest Studies in Philosophy, 10*(1), 615–678.

Block, N. (1987). Functional role and truth conditions. *Proceedings of the Aristotelian Society, Supplementary Volumes, 61*, 157–181.

Block, N. (2003). Do causal powers drain away? *Philosophy and Phenomenological Research, 67*(1), 133–150.

Bloom, P., & Keil, F. (2001). Thinking through language. *Mind & Language, 16*(4), 351–367.

Bohnemeyer, J. (2020). Linguistic relativity: From Whorf to now. In D. Gutzmann, L. Matthewson, C. Meier, H. Rullmann, and T. Zimmermann (eds.), *The Wiley Blackwell companion to semantics*. Wiley & Sons.

Bohon, C., Hembacher, E., Moller, H., Moody, T., & Feusner, J. (2012). Nonlinear relationships between anxiety and visual processing of own and others' faces in body dysmorphic disorder. *Psychiatry Research: Neuroimaging, 204*(2–3), 132–139.

Bonnet, L., Comte, A., Tatu, L., Millot, J. L., Moulin, T, & de Bustos, E. M. (2015). The role of the amygdala in the perception of positive emotions: An "intensity detector." *Frontiers in Behavioral Neuroscience, 9*, 178.

Boroditsky, L. (2012). How the languages we speak shape the ways we think: The FAQs. In M. J. Spivey, M. Joanisse, and K. McRae (eds.), *The Cambridge handbook of psycholinguistics* (pp. 615–632). Cambridge University Press.

Boroditsky, L., Schmidt, L. A., & Phillips, W. (2003). Sex, syntax, and semantics. In D. Gentner and S. Goldin-Meadow (eds.), *Language in mind: Advances in the study of language and thought* (pp. 61–79). MIT Press.

Boyd, R. (1989). What realism implies and what it does not. *Dialectica*, 43(1–2), 5–29.

Boyd, R. (1991). Realism, anti-foundationalism and the enthusiasm for natural kinds. *Philosophical Studies*, 61(1–2) 127–148.

Boyd, R. (2000). Kinds as the "workmanship of men": Realism, constructivism, and natural kinds. In J. Nida-Rümelin (ed.), *Rationality, realism, revision*, (pp. 52–89). Walter de Gruyter.

Boyer, P. (2008). Evolutionary economics of mental time travel? *Trends in Cognitive Sciences*, 12(6), 219–224.

Boyer, P. (2009). What are memories for? Functions of recall in cognition and culture. In P. Boyer and J. V. Wertsch (eds.), *Memory in mind and culture* (pp. 3–28). Cambridge University Press.

Boyer, P., & Barrett, H. C. (2005). Domain specificity and intuitive ontology. In D. Buss (ed.), *The handbook of evolutionary psychology* (pp. 96–118). Wiley.

Boyle, A. (2020). The impure phenomenology of episodic memory. *Mind & Language*, 35(5), 641–660.

Brigandt, I., & Griffiths, P. E. (2007). The importance of homology for biology and philosophy. *Biology & Philosophy*, 22(5), 633–641.

Bronstein, M. V., & Cannon, T. D. (2017). Bias against disconfirmatory evidence in a large nonclinical sample: Associations with schizotypy and delusional beliefs. *Journal of Experimental Psychopathology*, 8(3), 288–302.

Buhlmann, U., Etcoff, N. L., & Wilhelm, S. (2006). Emotion recognition bias for contempt and anger in body dysmorphic disorder. *Journal of Psychiatric Research*, 40(2), 105–111.

Buhlmann, U., Wacker, R., & Dziobek, I. (2015). Inferring other people's states of mind: Comparison across social anxiety, body dysmorphic, and obsessive-compulsive disorder. *Journal of Anxiety Disorders*, 34, 107–113.

Buhlmann, U., Winter, A., & Kathmann, N. (2013). Emotion recognition in body dysmorphic disorder: Application of the reading of the mind in eyes task. *Body Image*, 10(2), 247–250.

Burge, T. (1986). Individualism and psychology. *Philosophical Review*, 95(1), 3–45.

Burge, T. (2010). *Origins of objectivity*. Oxford University Press.

Camp, E. (2009). Putting thoughts to work: Concepts, systematicity, and stimulus-independence. *Philosophy and Phenomenological Research*, 78(2), 275–311.

Caramazza, A., & Mahon, B. Z. (2003). The organization of conceptual knowledge: The evidence from category-specific semantic deficits. *Trends in Cognitive Sciences*, 7(8), 354–361.

Caramazza, A., & Shelton, J. R. (1998). Domain-specific knowledge systems in the brain: The animate-inanimate distinction. *Journal of Cognitive Neuroscience*, 10(1), 1–34.

Carey, S. (1985). *Conceptual change in childhood*. MIT Press.

Carey, S. (1988). Conceptual differences between children and adults. *Mind and Language*, 3(3), 167–181.

Carey, S. (2009) *The origin of concepts*. Oxford University Press.

Carey, S., & Spelke, E. (1994). Domain-specific knowledge and conceptual change. In L. A. Hirschfeld and S. A. Gelman (eds.), *Mapping the mind: Domain specificity in cognition and culture* (pp. 169–200). Cambridge University Press.

Carroll, J. B. (Ed.) (1964). *Language, thought, and reality: Selected writings of Benjamin Lee Whorf.* MIT Press.

Carruthers, P. (2002). The cognitive functions of language. *Behavioral and Brain Sciences, 25*(6), 657–674.

Carruthers, P. (2012). Language in cognition. In E. Margolis, R. Samuels, and S. P. Stich (eds.), *Oxford handbook of philosophy of cognitive science* (pp. 382–401). Oxford University Press.

Casasanto, D. (2008). Who's afraid of the big bad Whorf? Crosslinguistic differences in temporal language and thought. *Language Learning, 58*(Suppl.1), 63–79.

Casasanto, D. (2016). Linguistic relativity. In N. Reimer (ed.), *Routledge handbook of semantics* (pp. 158–174). Routledge.

Casasanto, D., & Lupyan, G. (2015). All concepts are ad hoc concepts. In E. Margolis and S. Laurence (eds.), *The conceptual mind* (pp. 543–566). MIT Press.

Ceschi, A., Costantini, A., Sartori, R., Weller, J., & Di Fabio, A. (2019). Dimensions of decision-making: An evidence-based classification of heuristics and biases. *Personality and Individual Differences, 146*, 188–200.

Chalmers, D. (2008). Foreword. In A. Clark (ed.), *Supersizing the mind: Embodiment, action and cognitive extension* (pp. ix–xxix). Oxford University Press.

Charland, L. C. (2002). The natural kind status of emotion. *British Journal for the Philosophy of Science, 53*(4), 511–537.

Cheney, D. L., & Seyfarth, R. M. (1985). Social and non-social knowledge in vervet monkeys. *Philosophical Transactions of the Royal Society B, 308*(1135), 187–201.

Cheney, D. L., & Seyfarth, R. M. (1990). *How monkeys see the world: Inside the mind of another species.* University of Chicago Press.

Cheng, S., & Werning, M. (2016). What is episodic memory if it is a natural kind? *Synthese, 193*, 1345–1385.

Cheng, S., Werning, M., & Suddendorf, T. (2016). Dissociating memory traces and scenario construction in mental time travel. *Neuroscience and Biobehavioral Reviews, 60*, 82–89.

Chi, M. T. H., Feltovich, P. J., & Glaser, R. (1981). Categorization and representation of physics problems by experts and novices. *Cognitive Science, 5*(2), 121–152.

Chiandetti, C., & Vallortigara, G. (2008). Is there an innate geometric module? Effects of experience with angular geometric cues on spatial re-orientation based on the shape of the environment. *Animal Cognition, 11*(1), 139–146.

Child, W. (2001). Triangulation: Davidson, realism and natural kinds. *Dialectica, 55*(1), 29–49.

Churchland, P. M. (1982). Is "thinker" a natural kind? *Dialogue, 21*(2), 223–238.

Cibelli, E., Xu, Y., Austerweil, J. L., Griffiths, T. L., & Regier, T. (2016). The Sapir-Whorf hypothesis and probabilistic inference: Evidence from the domain of color. *PLoS ONE 11*(7), e0158725.

Cimpian, A., & Salomon, E. (2014). The inherence heuristic: An intuitive means of making sense of the world, and a potential precursor to psychological essentialism. *Behavioral and Brain Sciences, 37*(5), 461–480.

Clark, A. (1998). Magic words: How language augments human computation. In P. Carruthers and J. Boucher (eds.), *Language and thought: Interdisciplinary themes* (pp. 162–183). Cambridge University Press.
Clark, H. (1996). Communities, commonalities, and communication. In J. J. Gumperz and S. C. Levinson (eds.), *Rethinking linguistic relativity* (pp. 324–355). Cambridge University Press.
Close, J., Hahn, U., Hodgetts, C. J., & Pothos, E. M. (2010). Rules and similarity in adult concept learning. In D. Mareschal, P. C. Quinn, and S. E. G. Lea (eds.), *The making of human concepts* (pp. 29–51). Oxford University Press.
Collins, J. (2005). Nativism: In defense of a biological understanding. *Philosophical Psychology*, *18*(2), 157–177.
Connell, L., & Lynott, D. (2014). Principles of representation: Why you can't represent the same concept twice. *Topics in Cognitive Science*, *6*(3), 390–406.
Corns, J. (2012). *Pain is not a natural kind* (Unpublished doctoral dissertation). City University of New York.
Cosmides, L., & Tooby, J. (1994). Origins of domain specificity: The evolution of functional organization. In L. A. Hirschfeld and S. A. Gelman (eds.), *Mapping the mind: Domain specificity in cognition and culture* (pp. 84–116). Cambridge University Press.
Cowie, F. (1999). *What's within? Nativism reconsidered*. Oxford University Press.
Craver, C. F. (2007). *Explaining the brain: Mechanisms and the mosaic unity of neuroscience*. Oxford University Press.
Craver, C. F. (2009). Mechanisms and natural kinds. *Philosophical Psychology*, *22*(5), 575–594.
Craver, C. F. (2013). Functions and mechanisms: A perspectivalist view. In P. Huneman (ed.), *Functions: Selection and mechanisms* (pp. 133–158). Springer.
Craver, C. F. (2020). Remembering: Epistemic and empirical. *Review of Philosophy and Psychology*, *11*(2), 261–281.
Crick, F., & Koch, C. (1990). Towards a neurobiological theory of consciousness. *Seminars in Neuroscience*, *2*, 263–275.
Cronbach, L. J., & Meehl, P. E. (1955). Construct validity in psychological tests. *Psychological Bulletin*, *52*(4), 281.
Csibra, G. (2007). Teachers in the wild. *Trends in Cognitive Sciences*, *11*(3), 95–96.
Darden, L., & Maull, N. (1977). Interfield theories. *Philosophy of Science*, *44*(1), 43–64.
Davidson, D. (1973). Radical interpretation. *Dialectica*, *27*(3–4), 313–328.
Davidson, D. (1974). On the very idea of a conceptual scheme. *Proceedings and Addresses of the American Philosophical Association*, *47*, 5–20
De Brigard, F. (2014). Is memory for remembering? Recollection as a form of episodic hypothetical thinking. *Synthese*, *191*(2), 155–185.
De Brigard, F., Addis, D. R., Ford, J. H., Schacter, D. L., & Giovanello, K. S. (2013). Remembering what could have happened: Neural correlates of episodic counterfactual thinking. *Neuropsychologia*, *51*(12), 2401–2414.
Debus, D. (2016). Imagination and memory. In A. Kind (ed.), *The Routledge handbook of philosophy of imagination* (pp. 135–148). Routledge.

De Neys W. (2010) Heuristic bias, conflict, and rationality in decision-making. In B. Glatzeder, V. Goel, and A. Müller (eds.), *Towards a theory of thinking*, (pp. 22–33). Springer.

Dennett, D. C. (1987). *The intentional stance*. MIT Press.

Dennett, D. (1997). How to do other things with words. In J. Preston (ed.), *Thought and language* (pp. 219–235). Cambridge University Press.

Dèttore, D., & O'Connor, K. (2013). OCD and cognitive illusions. *Cognitive Therapy and Research*, 37(1), 109–121.

Dolscheid, S., Shayan, S., Majid, A., & Casasanto, D. (2013). The thickness of musical pitch: Psychophysical evidence for linguistic relativity. *Psychological Science*, 24(5), 613–621.

Earman, J. (1977). Against indeterminacy. *Journal of Philosophy*, 74(9), 535–538.

Earman, J., & Friedman, M. (1973). The meaning and status of Newton's law of inertia and the nature of gravitational forces. *Philosophy of Science*, 40(3), 329–359.

Edwards, K., & Smith, E. E. (1996). A disconfirmation bias in the evaluation of arguments. *Journal of Personality and Social Psychology*, 71(1), 5–24.

Egan, F. (2014). How to think about mental content. *Philosophical Studies*, 170(1), 115–135.

Eisen, J. L., Phillips, K. A., Coles, M. E., & Rasmussen, S. A. (2004). Insight in obsessive compulsive disorder and body dysmorphic disorder. *Comprehensive Psychiatry*, 45(1), 10–15.

Eichenbaum, H. (2007). Persistence: Necessary, but not sufficient. In H. Roediger, Y. Dudai, and S. Fitzpatrick (eds.), *Science of memory: Concepts* (pp. 193–198). Oxford University Press.

Epstein, B. (2009). Ontological individualism reconsidered. *Synthese*, 166(1), 187–213.

Epstein, B. (2015). *The ant trap: Rebuilding the foundations of the social sciences*. Oxford University Press.

Ereshefsky, M. (2007). Psychological categories as homologies: Lessons from ethology. *Biology & Philosophy*, 22(5), 659–674.

Ereshefsky, M. (2012). Homology thinking. *Biology & Philosophy*, 27(3), 381–400.

Evans, J. S. B. T. (1989). *Bias in human reasoning: Causes and consequences*. Lawrence Erlbaum Associates, Inc.

Evans, J. S. B. T. (2011). Dual-process theories of reasoning: Contemporary issues and developmental applications. *Developmental Review*, 31(2–3), 86–102.

Evans, J. S. B. T. (2012). Questions and challenges for the new psychology of reasoning. *Thinking & Reasoning*, 18(1), 5–31.

Evans, J. S. B. T., & Stanovich, K. E. (2013). Dual-process theories of higher cognition: Advancing the debate. *Perspectives on Psychological Science*, 8(3), 223–241.

Feusner, J., Hembacher, E., Moller, H., & Moody, T. (2011). Abnormalities of visual processing in body dysmorphic disorder. *Psychological Medicine*, 41(11), 2385–2397.

Feusner, J., Neziroglu, F., Wilhelm, S., Mancusi, L., & Bohon, C. (2010). What causes BDD: Research findings and a proposed model. *Psychiatric Annals*, 40(7), 349–355.

Feusner, J. D., Moller, H., Altstein, L., Sugar, C., Bookheimer, S., Yoon, J., & Hembacher, E. (2010). Inverted face processing in body dysmorphic disorder. *Journal of Psychiatric Research, 44*(15), 1088–1094.

Fine, C., Gardner, M., Craigie, J., & Gold, I. (2007). Hopping, skipping or jumping to conclusions? Clarifying the role of the JTC bias in delusions. *Cognitive Neuropsychiatry, 12*(1), 46–77.

Flake, J. K., & Fried, E. I. (2020). Measurement schmeasurement: Questionable measurement practices and how to avoid them. *Advances in Methods and Practices in Psychological Science, 3*(4), 456–465.

Fodor, J. A. (1981). *Representations: Philosophical essays on the foundations of cognitive science*. MIT Press.

Fodor, J. A. (1983). *The modularity of mind*. MIT Press.

Fodor, J. A. (1987). *Psychosemantics: The problem of meaning in the philosophy of mind*. MIT Press.

Fodor, J. A. (1998). *Concepts: Where cognitive science went wrong*. Oxford University Press.

Fodor, J. A. (2000). *The mind doesn't work that way*. MIT Press.

Fodor, J. A., Garrett, M. F., Walker, E. C. T., & Parkes, C. H. (1980). Against definitions. *Cognition, 8*(3), 263–367.

Foster, J. K. (2009). *Memory: A very short introduction*. Oxford University Press.

Francken, J. C., & Slors, M. (2014). From commonsense to science, and back: The use of cognitive concepts in neuroscience. *Consciousness and Cognition, 29*, 248–258.

Francken, J. C., & Slors, M. (2018). Neuroscience and everyday life: Facing the translation problem. *Brain and Cognition, 120*, 67–74.

Garety, P. A., Bebbington, P., Fowler, D., Freeman, D., & Kuipers, E. (2007). Implications for neurobiological research of cognitive models of psychosis: A theoretical paper. *Psychological Medicine, 37*(10), 1377–1391.

Garson, J. (2011). Selected effects and causal role functions in the brain: The case for an etiological approach to neuroscience. *Biology and Philosophy, 26*(4), 547–565.

Garson, J. (2019). *What biological functions are and why they matter*. Cambridge University Press.

Gelman, S. A. (2004). Psychological essentialism in children. *Trends in Cognitive Sciences, 8*(9), 404–409.

Gennari, S. P., Sloman, S. A., Malt, B. C., & Fitch, W. T. (2002). Motion events in language and cognition. *Cognition, 83*(1), 49–79.

Gentner, D. (2003). Why we're so smart. In D. Gentner and S. Goldin-Meadow (eds.), *Language in mind* (pp. 195–235). MIT Press.

Gigerenzer, G. (2004). Fast and frugal heuristics: The tools of bounded rationality. In D. Koehler and N. Harvey (eds.), *Blackwell handbook of judgment and decision making* (pp. 19–36). Blackwell.

Gigerenzer, G. (2018). The bias bias in behavioral economics. *Review of Behavioral Economics, 5*(3–4), 303–336.

Gigerenzer, G., & Brighton, H. (2009). Homo heuristicus: Why biased minds make better inferences. *Topics in Cognitive Science, 1*(1), 107–143.

Gilovich, T., & Griffin, D. (2002). Introduction – Heuristics and biases: Then and now. In T. Gilovich, D. Griffin, and D. Kahneman, (eds.), *Heuristics and biases: The psychology of intuitive judgment* (pp. 1–18). Cambridge University Press.
Gleitman, L. R., Armstrong, S. L., & Gleitman, H. (1983). On doubting the concept "concept." In E. Kofsky Scholnick (ed.), *New trends in conceptual representation: Challenges to Piaget's theory?* (pp. 87–110). Lawrence Erlbaum.
Gleitman, L., & Papafragou, A. (2005). Language and thought. In K. J. Holyoak and R. G. Morrison (eds.), *The Cambridge handbook of thinking and reasoning* (pp. 633–661). Cambridge University Press.
Glennan, S. (2017). *The new mechanical philosophy*. Oxford University Press.
Godfrey-Smith, P. (2010). Causal pluralism. In H. Beebee, C. Hitchcock, and P. Menzies (eds.), *Oxford handbook of causation* (pp. 326–337). Oxford University Press.
Goodman, N. (1954/1979). *Fact, fiction, and forecast*. Harvard University Press.
Gopnik, A. (1984). Conceptual and semantic change in scientists and children: Why there are no semantic universals. *Linguistics, 20*, 163–179.
Gopnik, A. (1988). Conceptual and semantic development as theory change: The case of object permanence. *Mind & Language, 3*(3), 197–216.
Gottfried, G. M., & Gelman, S. A. (2005). Developing domain-specific causal-explanatory frameworks: The role of insides and immanence. *Cognitive Development, 20*(1), 137–158.
Grace, S. A., Toh, W. L., Buchanan, B., Castle, D. J., & Rossell, S. L. (2019). Impaired recognition of negative facial emotions in body dysmorphic disorder. *Journal of the International Neuropsychological Society, 25*(8), 884–889.
Greenberg, G. (2013). *The book of woe: The DSM and the unmaking of psychiatry*. Blue Rider Press.
Greif, M. L., Kemler Nelson, D. G., Keil, F. C., & Gutierrez, F. (2006). What do children want to know about animals and artifacts? Domain-specific requests for information. *Psychological Science, 17*(6), 455–459.
Griffiths, P. E. (1993). Functional analysis and proper functions. *British Journal for the Philosophy of Science, 44*(3), 409–422.
Griffiths, P. E. (2002). What is innateness? *Monist, 85*(1), 70–85.
Griffiths, P. E. (2004). Is emotion a natural kind? In R. C. Solomon (ed.), *Thinking about feeling: Contemporary philosophers on emotions* (pp. 233–249). Oxford University Press.
Griffiths P. E., & Machery, E. (2008). Innateness, canalization and "biologicizing the mind." *Philosophical Psychology, 21*(3), 397–414.
Griffiths, P. E., & Stotz, K. (2008). Experimental philosophy of science. *Philosophy Compass, 3*(3), 507–521.
Griffiths, P. E., Machery, E., & Linquist, S. (2009). The vernacular concept of innateness. *Mind & Language, 24*(5), 605–630.
Gross, C. G. (2002). Genealogy of the "grandmother cell." *The Neuroscientist, 8*(5), 512–518.
Hacking, I. (1991). A tradition of natural kinds. *Philosophical Studies, 61*(1–2), 109–126.

Hacking, I. (1995). The looping effects of human kinds. In D. Sperber, D. Premack, and A. J. Premack (eds.), *Causal cognition: A multidisciplinary debate* (pp. 351–394). Oxford University Press.

Hahn, U., & Harris, A. J. L. (2014). What does it mean to be biased: Motivated reasoning and rationality. In B. H. Ross (ed.), *The psychology of learning and motivation: Vol. 61. The psychology of learning and motivation* (pp. 41–102). Elsevier Academic Press.

Haidt, J., & Joseph, C. (2007) The moral mind: How five sets of innate moral intuitions guide the development of many culture-specific virtues, and perhaps even modules. In P. Carruthers, S. Laurence, and S. Stich (eds.), *The innate mind* (pp. 367–392). Oxford University Press.

Hampton, J. A. (1997). Conceptual combination. In K. Lamberts and D. R. Shanks (eds.), *Knowledge, concepts and categories* (pp. 135–162). Psychology Press.

Hampton, J. A. (2006). Concepts as prototypes. In B. H. Ross (ed.), *The psychology of learning and motivation: Advances in research and theory* (pp. 79–113). Elsevier Academic Press.

Harman, G. (1982). Conceptual role semantics. *Notre Dame Journal of Formal Logic*, *23*(2), 242–256.

Harris, C., Rasmussen, A., & Berntsen, D. (2014). The functions of autobiographical memory: An integrative approach. *Memory*, *22*(5), 559–581.

Haslanger, S. (2000). Gender and race: (What) are they? (What) do we want them to be? *Noûs*, *34*(1), 31–55.

Hassabis, D., Kumaran, D., Vann, S. D., & Maguire, E. A. (2007). Patients with hippocampal amnesia cannot imagine new experiences. *Proceedings of the National Academy of Sciences*, *104*(5), 1726–1731.

Haugeland, J. (1998). Mind embodied and embedded. In J. Haugeland (ed.), *Having thought: Essays in the metaphysics of mind* (pp. 207–237). Harvard University Press.

Hauk, O., Johnsrude, I., & Pulvermüller, F. (2004). Somatotopic representation of action words in the motor and premotor cortex. *Neuron*, *41*(2), 301–307.

Heil, J. (1978). Traces of things past. *Philosophy of Science*, *45*(1), 60–72.

Held, B. (2017). The distinction between psychological kinds and natural kinds revisited: Can updated natural-kind theory help clinical psychological science and beyond meet psychology's philosophical challenges? *Review of General Psychology*, *21*(1), 82–94.

Hermer, L., & Spelke, E. (1996). Modularity and development: The case of spatial reorientation. *Cognition*, *61*(3), 195–232.

Hezel, D. M., & McNally, R. J. (2016). A theoretical review of cognitive biases and deficits in obsessive–compulsive disorder. *Biological Psychology*, *121*(pt. B), 221–232.

Hirschfeld, L. A., & Gelman, S. A. (1994). Toward a topography of mind: An introduction to domain specificity. In L. A. Hirschfeld and S. A. Gelman (eds.), *Mapping the mind: Domain specificity in cognition and culture* (pp. 3–35). Cambridge University Press.

Hoenig, K., Muller, C., Herrnberger, B., Sim, E. J., Spitzer, M., Ehret, G., & Kiefer, M. (2011). Neuroplasticity of semantic representations for musical instruments in professional musicians. *Neuroimage, 56*(3), 1714–1725.

Hoenig, K., Sim, E.-J., Bochev, V., Herrnberger, B., & Kiefer, M. (2008) Conceptual flexibility in the human brain: Dynamic recruitment of semantic maps from visual, motion and motor-related areas. *Journal of Cognitive Neuroscience, 20*(10), 1799–1814.

Hoerl, C. (2001). The phenomenology of episodic recall. In C. Hoerl and T. McCormack (eds.), *Time and memory: Issues in philosophy and psychology* (pp. 315–338). Oxford University Press.

Hoerl, C. & McCormack, T. (2016). Making decisions about the future. In K. Michaelian, S. B. Klein, and K. K. Szpunar (eds.), *Seeing the future: Theoretical perspectives on future-oriented mental time travel* (pp. 241–266). Oxford University Press.

Holmes, K. J., & Wolff, P. (2010). Simulation from schematics: Dorsal stream processing and the perception of implied motion. *Proceedings of the 32nd Annual Conference of the Cognitive Science Society* (pp. 2704–2709). Cognitive Science Society.

Hunt, E., & Agnoli, F. (1991). The Whorfian hypothesis: A cognitive psychology perspective. *Psychological Review, 98*(3), 377–389.

Hutto, D. D., Peeters, A., & Segundo-Ortin, M., (2017). Cognitive ontology in flux: The possibility of protean brains. *Philosophical Explorations, 20*(2), 209–223.

Imai, M., & Gentner, D. (1997). A cross-linguistic study of early word meaning: Universal ontology and linguistic influence. *Cognition, 62*(2), 169–200.

Irish, M., Addis, D. R., Hodges, J. R., & Piguet, O. (2012). Considering the role of semantic memory in episodic future thinking: evidence from semantic dementia. *Brain, 135*(7), 2178–2191.

Kahneman, D. (2011). *Thinking fast and slow*. Farrar, Strauss & Giroux.

Kahneman, D., & Frederick, S. (2002). Representativeness revisited: Attribute substitution in intuitive judgment. In T. Gilovich, D. Griffin, and D. Kahneman (eds.), *Heuristics and biases: The psychology of intuitive judgment* (pp. 49–81). Cambridge University Press.

Kaplan, R., Rossell, S., Enticott, P., & Castle, D. (2013). Own-body perception in body dysmorphic disorder. *Cognitive Neuropsychiatry, 18*(6), 594–614.

Kay, P., & Kempton, W. (1984). What is the Sapir-Whorf hypothesis? *American Anthropologist, 86*(1), 65–79.

Keil, F.C. (1986). The acquisition of natural kind and artifact terms. In W. Demopoulos and A. Marras (eds.), *Language learning and concept acquisition: Foundational issues* (pp. 133–153). Ablex.

Keil, F.C. (1989). *Concepts, kinds, and cognitive development*. MIT Press.

Keil, F. C. (1996). The growth of causal understandings of natural kinds. In Sperber, D., Premack, D. and Premack, A. J. (eds.), *Causal cognition: A multidisciplinary debate* (pp. 235–267). Oxford University Press.

Kemmerer, D. L. (2019). *Concepts in the brain: The view from cross-linguistic diversity*. Oxford University Press.

Kendig, C. (2015). Homologizing as kinding. In Kendig, C. (ed.), *Natural kinds and classification in scientific practice* (pp. 126–146). Routledge.
Khalidi, M. A. (1995). Two concepts of concept. *Mind and Language*, *10*(4), 402–422.
Khalidi, M. A. (1998). Incommensurability in cognitive guise. *Philosophical Psychology*, *11*(1), 29–43.
Khalidi, M. A. (2001). Innateness and domain specificity. *Philosophical Studies*, *105*(2), 191–210.
Khalidi, M. A. (2002). Nature and nurture in cognition. *British Journal for the Philosophy of Science*, *53*(2), 251–272.
Khalidi, M. A. (2007). Innate cognitive capacities. *Mind and Language*, *22*(1), 92–115.
Khalidi, M. A. (2011). The pitfalls of microphysical realism. *Philosophy of Science*, *78*(5), 1156–1164.
Khalidi, M. A. (2013). *Natural categories and human kinds: Classification in the natural and social sciences*. Cambridge University Press.
Khalidi, M. A. (2016a). Innateness as a natural cognitive kind. *Philosophical Psychology*, *29*(3), 319–333.
Khalidi, M. A. (2016b). Mind-dependent kinds. *Journal of Social Ontology*, *2*(2) 223–246.
Khalidi, M. A. (2017). Crosscutting psycho-neural taxonomies: The case of episodic memory. *Philosophical Explorations*, *20*(2), 191–208.
Khalidi, M. A. (2018). Natural kinds as nodes in causal networks. *Synthese*, *195*(4), 1379–1396.
Khalidi, M. A. (2020). Neural correlates without reduction: The case of the critical period. *Synthese*, *197*(5), 1947–1959.
Khalidi, M. A. (2021). Etiological kinds. *Philosophy of Science*, *88*(1), 1–21.
Kim, J. (1992). Multiple realization and the metaphysics of reduction. *Philosophy and Phenomenological Research*, *52*(1), 1–26.
Kincaid, H., & Sullivan, J. A. (Eds.). (2014). *Classifying psychopathology: Mental kinds and natural kinds*. MIT Press.
Kitcher, P. (1984). Species. *Philosophy of Science*, *51*(2), 308–333.
Kitcher, P. (1997). *The lives to come: The genetic revolution and human possibilities*. Free Press.
Kitcher, P. (1998). Marr's computational theory of vision. *Philosophy of Science*, *55*(1), 1–24.
Klayman, J. (1995). Varieties of confirmation bias. In J. Busemeyer, R. Hastie, & D. L. Medin (eds.), *Psychology of learning and motivation: Vol. 32* (pp. 385–418). Academic Press.
Klayman, J., & Ha, Y. W. (1987). Confirmation, disconfirmation, and information in hypothesis testing. *Psychological Review*, *94*(2), 211–228.
Klein, C. (2012). Cognitive ontology and region- versus network-oriented analyses. *Philosophy of Science*, *79*(5), 952–960.
Klein, S. B. (2013). Making the case that episodic recollection is attributable to operations occurring at retrieval rather than to content stored in a dedicated subsystem of long-term memory. *Frontiers in Behavioral Neuroscience*, *7*(3), 1–14.

Klein, S. B. (2015). What memory is. *WIREs Cognitive Science*, *6*(1), 1–38.

Klein, S. B., Cosmides, L., Tooby, J., & Chance, S. (2002). Decisions and the evolution of memory: Multiple systems, multiple functions. *Psychological Review*, *109*(2), 306–329.

Knobe, J., & Samuels, R. (2013). Thinking like a scientist: Innateness as a case study. *Cognition*, *126*(1), 72–86.

Koriat, A., & Goldsmith, M. (1996). Monitoring and control processes in the strategic regulation of memory accuracy. *Psychological Review*, *103*(3), 490–517.

Kornblith, H. (1993). *Inductive inference and its natural ground*. MIT Press.

Kornblith, H. (1994). Naturalism: Both metaphysical and epistemological. *Midwest Studies in Philosophy*, *19*(1), 39–52 (reprinted in Kornblith, H. (2014). *A naturalistic epistemology*. Oxford: Oxford University Press.)

Kripke, S. (1980). *Naming and necessity*. Harvard University Press.

Kruglanski, A. W., & Gigerenzer, G. (2011). Intuitive and deliberate judgments are based on common principles. *Psychological Review*, *118*(1), 97–109.

Kunda, Z. (1990). The case for motivated reasoning. *Psychological Bulletin*, *108*(3), 480–498.

Kwan, D., Craver, C. F., Green, L., Myerson, J., Boyer, P., & Rosenbaum, S. (2012). Future decision-making without episodic mental time travel. *Hippocampus*, *22*(6), 1215–1219.

Kwan, D., Craver, C. F., Green, L., Myerson, J., Gao, F., Black, S. E., & Rosenbaum, R. S. (2015). Cueing the personal future to reduce discounting in intertemporal choice: Is episodic prospection necessary? *Hippocampus*, *25*(4), 432–443.

Ladyman, J., & Ross, D. (2007). *Every thing must go: Metaphysics naturalized*. Oxford University Press.

Laurence, S., & Margolis, E. (1999). Concepts and cognitive science. In E. Margolis & S. Laurence (eds.), *Concepts: CoreReadings* (pp. 3–81). MIT Press.

Laurence, S., & Margolis, E. (2001). The poverty of the stimulus argument. *British Journal for the Philosophy of Science*, *52*(2), 217–276.

Lebois, L. A. M., Wilson-Mendenhall, C. D., & Barsalou, L. W. (2015). Are automatic conceptual cores the gold standard of semantic processing? The context-dependence of spatial meaning in grounded congruency effects. *Cognitive Science*, *39*(8): 1764–1801.

Lello, L., Avery, S. G., Tellier, L., Vazquez, A. I., Campos, G., & Hsu, S. D. H. (2018). Accurate genomic prediction of human height. *Genetics*, *210*(2), 477–497.

Leslie, S. J. (2013). Essence and natural kinds: When science meets preschooler intuition. *Oxford Studies in Epistemology*, *4*, 108–166.

Levinson, S. C. (2003). *Space in language and cognition: Explorations in cognitive diversity*. Cambridge University Press.

Lewis, D. (1983). New work for a theory of universals. *Australasian Journal of Philosophy*, *61*(4), 343–377.

Li, P., Dunham, Y., & Carey, S. (2009). Of substance: The nature of language effects on entity construal. *Cognitive Psychology*, *58*(4), 487–524.

Linquist, S., Machery, E., Griffiths, P. E., & Stotz, K. (2011). Exploring the folk-biological conception of human nature. *Philosophical Transactions of the Royal Society B*, *366*(1563), 444–453.

Lindquist, K., Wager, T., Kober, H., Bliss-Moreau, E., & Barrett, L. (2012). The brain basis of emotion: A meta-analytic review. *Behavioral and Brain Sciences*, *35*(3), 121–143.

Lisman, J. (2007). Integrative comments persistence: In search of molecular persistence. In H. Roediger, Y. Dudai, and S. Fitzpatrick (eds.), *Science of memory: Concepts* (pp. 203–208). Oxford University Press.

Liu, S., Brooks, N. B., & Spelke, E. S. (2019). Origins of the concepts cause, cost, and goal in prereaching infants. *Proceedings of the National Academy of Sciences of the United States of America*, *116*(36), 17747–17752.

Liu, X., Ramirez, S., & Tonegawa, S. (2014). Inception of a false memory by optogenetic manipulation of a hippocampal memory engram. *Philosophical Transactions of the Royal Society of London. Series B, Biological sciences*, *369*(1633), 20130142.

Loftus, E. (2003). Our changeable memories: Legal and practical implications. *Nature Reviews Neuroscience*, *4*(3), 231–234.

Loftus, E. F., & Pickrell, J. E. (1995). The formation of false memories. *Psychiatric annals*, *25*(12), 720–725.

Lord, C. G., Ross, L., & Lepper, M. R. (1979). Biased assimilation and attitude polarization: The effects of prior theories on subsequently considered evidence. *Journal of Personality and Social Psychology*, *37*(11), 2098–2109.

Lowe, E. J. (2006). *The four-category ontology: A metaphysical foundation for natural science*. Oxford University Press.

Ludwig, D. (2016). Ontological choices and the value-free ideal. *Erkenntnis*, *81*, 1253–1272.

Lupyan, G. (2012). Linguistically modulated perception and cognition: The label-feedback hypothesis. *Frontiers in Psychology*, *3*(54), 1–13.

Lupyan, G., & Thompson-Schill, S. L. (2012). The evocative power of words: Activation of concepts by verbal and nonverbal means. *Journal of Experimental Psychology, General*, *141*(1), 170–186.

Lupyan, G., Rahman, R. A., Boroditsky, L., & Clark, A. (2020). Effects of language on visual perception. *Trends in Cognitive Sciences*, *24*(11), 930–944.

Machamer, P., Darden, L., & Craver, C. F. (2000). Thinking about mechanisms. *Philosophy of Science*, *67*(1), 1–25.

Machery, E. (2005). Concepts are not a natural kind. *Philosophy of Science*, *72*(3), 444–467.

Machery, E. (2007). Concept empiricism: A methodological critique. *Cognition*, *104*(1), 19–46.

Machery, E. (2009). *Doing without concepts*. Oxford University Press.

Maher, B. A. (1988). Anomalous experience and delusional thinking: The logic of explanations. In T. F. Oltmanns and B. A. Maher (eds.), *Delusional beliefs* (pp. 15–33). Wiley.

Mahon, B. Z., & Hickok, G. (2016). Arguments about the nature of concepts: Symbols, embodiment, and beyond. *Psychonomic Bulletin & Review*, *23*(4), 941–958.

Mahr, J. B., & Csibra, G. (2018). Why do we remember? The communicative function of episodic memory. *Behavioral and Brain Sciences*, 41, 1–16.

Majid, A. (2002). Frames of reference and language concepts. *Trends in Cognitive Sciences*, 6(12), 503–504.

Majid, A., Bowerman, M., Kita, S., Haun, D. B. M., & Levinson, S. C. (2004). Can language restructure cognition? The case for space. *Trends in Cognitive Sciences*, 8(3), 108–114.

Mallon, R. (2003). Social construction, social roles, and stability. In F. Schmitt (ed.), *Socializing metaphysics: The nature of social reality* (pp. 327–354). Rowman & Littlefield.

Mallon, R., & Weinberg, J. M. (2006). Innateness as closed process invariance. *Philosophy of Science*, 73(3), 323–344.

Mameli, M. (2008). On innateness: The clutter hypothesis and the cluster hypothesis. *Journal of Philosophy*, 105(12), 719–736.

Mameli, M., & Bateson, P. E. (2006). Innateness and the sciences. *Biology and Philosophy*, 21, 155–188.

Mameli, M., & Bateson, P. E. (2011). An evaluation of the concept of innateness. *Philosophical Transactions of the Royal Society B*, 366(1563), 436–443.

Mandelbaum, D. (Ed.) (1949). *Selected writings of Edward Sapir*. University of California Press.

Mandler, J. M. (1992). How to build a baby: II. Conceptual primitives. *Psychological Review*, 99(4), 587–604.

Marr, D. (1982). *Vision: A computational investigation into the human representation and processing of visual information*. W.H. Freeman.

Martin, C. B., & Deutscher, M. (1966). Remembering. *The Philosophical Review*, 75(2), 161–196.

McCaffrey, J. (2015a). Reconceiving conceptual vehicles: Lessons from semantic dementia. *Philosophical Psychology*, 28(3), 337–354.

McCaffrey, J. (2015b). The brain's heterogeneous functional landscape. *Philosophy of Science*, 82(5), 1010–1022.

McCaffrey, J. & Machery, E. (2012). Philosophical issues about concepts. *Wiley Interdisciplinary Reviews*, 3(2), 265–279.

McKone, E., Kanwisher, N., & Duchaine, B. C. (2007). Can generic expertise explain special processing for faces? *Trends in Cognitive Sciences*, 11(1), 8–15.

McGeer, V. (2015). Mind-making practices: The social infrastructure of self-knowing agency and responsibility. *Philosophical Explorations*, 18(2), 259–281.

McIntosh, A. R. (2004). Contexts and catalysts: A resolution of the localization and integration of function in the brain. *Neuroinformatics*, 2(2), 175–182.

McKone, E. Kanwisher, N., & Duchaine, B.C. (2007). Can generic expertise explain special processing for faces? *Trends in Cognitive Sciences*, 11(1), 8–15.

McLean, B., Mattiske, J., & Balzan, R. (2017). Association of the jumping to conclusions and evidence integration biases with delusions in psychosis: A detailed meta-analysis. *Schizophrenia Bulletin*, 43(2), 344–354.

Mercier, H. (2017). Confirmation bias: Myside bias. In R. F. Pohl (ed.), *Cognitive illusions: Intriguing phenomena in thinking, judgment and memory* (pp. 99–114). Routledge/Taylor & Francis Group.

Mercier, H., & Sperber, D. (2011). Why do humans reason? Arguments for an argumentative theory. *Behavioral and Brain Sciences*, 34(2), 57–74.

Michaelian, K. (2011a). Generative memory. *Philosophical Psychology*, 24(3), 323–342.

Michaelian, K. (2011b). Is memory a natural kind? *Memory Studies*, 4(2), 170–189.

Michaelian, K. (2016). *Mental time travel: Episodic memory and our knowledge of the personal past*. MIT Press.

Mill, J. S. (1843/1882). *A system of logic* (8th edition). Harper & Brothers.

Millikan, R. G. (1989). In defense of proper functions. *Philosophy of Science*, 56(2), 288–302.

Millikan, R. G. (1999). Historical kinds and the "special sciences." *Philosophical Studies*, 95(1/2), 45–65.

Morris, R. G. M. (2007). Integrative comments: Distinctions and dilemmas. In H. L. Roediger, Y. Dudai, and S. M. Fitzpatrick (eds.), *Science of memory: Concepts* (pp. 29–34). Oxford University Press.

Mufaddel, A., Osman, O. T., Almugaddam, F., & Jafferany, M. (2013). A review of body dysmorphic disorder and its presentation in different clinical settings. *The Primary Care Companion for CNS Disorders*, 15(4), PCC.12r01464.

Mugg, J. (2016). The dual-process turn: How recent defenses of dual-process theories of reasoning fail. *Philosophical Psychology*, 29(2), 300–309.

Mugg, J., & Khalidi, M. A. (2021). Self-reflexive cognitive bias. *European Journal for Philosophy of Science*, 11(3), 1–21.

Mundy, M., & Sadusky, A. (2014). Abnormalities in visual processing amongst students with body image concerns. *Advances in Cognitive Psychology*, 10(2), 39–48.

Munro, G. D., & Stansbury, J. A. (2009). The dark side of self-affirmation: Confirmation bias and illusory correlation in response to threatening information. *Personality and Social Psychology Bulletin*, 35(9), 1143–1153.

Murphy, D. (2006). *Psychiatry in the scientific image*. MIT Press.

Murphy, G. L., & Medin, D. L. (1985) The role of theories in conceptual coherence. *Psychological Review*, 92(3), 289–316.

Nadel, L., Samsonovich, A., Ryan, L., & Moscovitch, M. (2000). Multiple trace theory of human memory: computational, neuroimaging, and neuropsychological results. *Hippocampus*, 10(4), 352–368.

Nader, K., & Hardt, O. (2009). A single standard for memory: The case for reconsolidation. *Nature Reviews Neuroscience*, 10(3), 224–234.

Nagel, E. (1961). *The structure of science: Problems in the logic of scientific explanation*. Harcourt, Brace & World.

Nathan, M. J., & Del Pinal, G. D. (2016). Mapping the mind: Bridge laws and the psycho-neural interface. *Synthese*, 193(2), 637–657.

Neander, K. (1991). Functions as selected effects: The conceptual analyst's defense. *Philosophy of Science*, 58(2), 168–184.

Newell, A., & Simon, H. A. (1976). Computer science as empirical inquiry: Symbols and search. *Communications of the ACM, 19*(3), 113–126.

Newell, A., & Simon, H. A. (2007). Computer science as empirical inquiry: Symbols and search. *ACM Turing Award Lectures, 19*(3), 113–126.

Nickerson, R. S. (1996). Hempel's paradox and Wason's selection task: Logical and psychological puzzles of confirmation. *Thinking and Reasoning, 2*(1), 1–31.

Nickerson, R. S. (1998). Confirmation bias: A ubiquitous phenomenon in many guises. *Review of General Psychology, 2*(2), 175–220.

Nyhan, B., & Reifler, J. (2010). When corrections fail: The persistence of political misperceptions. *Political Behavior, 32*(2), 303–330.

Oldham, J., Hollander, E., & Skodol, A. (Eds.) (1996). *Impulsivity and compulsivity*. American Psychiatric Press.

Oosthuizen, P., Lambert, T., & Castle, D. J. (1998). Dysmorphic concern: Prevalence and associations with clinical variables. *Australian and New Zealand Journal of Psychiatry, 32*(1), 129–132.

Oppenheim, P., & Putnam, H. (1958). Unity of science as a working hypothesis. *Minnesota Studies in the Philosophy of Science, 2*, 3–36.

Oreg, S., & Bayazit, M. (2009). Prone to bias: Development of a bias taxonomy from an individual differences perspective. *Review of General Psychology, 13*(3), 175–193.

Oswald, M. E., & Grosjean, S. (2004). Confirmation bias. In R. F. Pohl (ed.), *Cognitive illusions: A handbook on fallacies and biases in thinking, judgment, and memory*. Psychology Press.

Partanen, E., Kujala, T., Tervaniemi, M., & Huotilainen, M. (2013). Prenatal music exposure induces long-term neural effects. *PLoS ONE, 8*(10), e78946.

Perry, L. K., & Lupyan, G. (2013). What the online manipulation of linguistic activity can tell us about language and thought. *Frontiers in Behavioral Neuroscience, 7*, 122.

Phillips, K. (1996/2005). *The broken mirror: Understanding and treating body dysmorphic disorder* (2nd edition). Oxford University Press.

Phillips, K., Pinto, A., Hart, A., Coles, M., Eisen, J., Menard, W., & Rasmmussen, S. (2012). A comparison of insight in body dysmorphic disorder and obsessive compulsive disorder. *Journal of Psychiatric Research, 46*(10), 1293–1299.

Phillips, K. A. (1996/2005). *The broken mirror: Understanding and treating Body Dysmorphic Disorder*. Oxford University Press.

Phillips, K. A., Wilhelm, S., Koran, L. M., Didie, E. R., Fallon, B. A., Feusner, J., & Stein, D. J. (2010). Body dysmorphic disorder: Some key issues for DSM-V. *Depression and anxiety, 27*(6), 573–591.

PhilPapers Survey (2009). https://philpapers.org/surveys/results.pl (accessed March 9, 2020).

Pinker, S. (1995). *The Language instinct: How the mind creates language*. Harper.

Pober, J. M. (2013). Addiction is not a natural kind. *Frontiers in Psychiatry, 4*, 1–11.

Pohl, R. F. (2017). Cognitive illusions. In R. F. Pohl (ed.), *Cognitive illusions: Intriguing phenomena in thinking, judgment and memory* (2nd edition) (pp. 3–21). Routledge.

Poldrack, R. A. (2012). The future of fMRI in cognitive neuroscience. *Neuroimage*, *62*(2), 1216–1220.

Poldrack, R. A., Halchenko, Y. O., & Hanson, S. J. (2009). Decoding the large-scale structure of brain function by classifying mental states across individuals. *Psychological Science*, *20*(11), 1364–1372.

Poldrack, R. A., & Yarkoni, T. (2016). From brain maps to cognitive ontologies: Informatics and the search for mental structure. *Annual Review of Psychology*, *67*(1), 587–612.

Polger, T. W. (2002). Putnam's intuition. *Philosophical Studies*, *109*(2), 143–170.

Polger, T. W. (2009). Evaluating the evidence for multiple realization. *Synthese*, *167*(3), 457–472.

Polimeni, J., & Reiss, J. P. (2003). Evolutionary perspectives on schizophrenia. *The Canadian Journal of Psychiatry*, *48*(1), 34–39.

Pöyhönen, S. (2013). Carving the mind by its joints: Culture-bound psychiatric disorders as natural kinds. In M. Miłkowskiand K. Talmont-Kaminski (eds.), *Regarding the mind, naturally: Naturalist approaches to the sciences of the mental* (pp. 30–48). Cambridge Scholars Publishing.

Pöyhönen, H. S. (2015). Memory as a cognitive kind: Brains, remembering dyads, and exograms. In C. Kendig (ed.), *Natural kinds and classification in scientific practice* (pp. 145–156). Routledge/Taylor & Francis Group.

Price, C., & Friston, K. (2005) Functional ontologies for cognition: The systematic definition of structure and function. *Cognitive Neuropsychology*, *22*(3–4), 262–275.

Putnam, H. (1974). Meaning and reference. *Journal of Philosophy*, *70*(19), 699–711.

Putnam, H. (1975). *Mind, language and reality*. Cambridge University Press.

Quian Quiroga, R., Reddy, L., Kreiman, G., Koch, C., & Fried, I. (2005). Invariant visual representation by single neurons in the human brain. *Nature*, *435*(7045), 1102–1107.

Rabin, G. O. (2018). Grounding orthodoxy and the layered conception. In R. Bliss and G. Priest (eds.), *Reality and its structure: Essays in fundamentality* (pp. 37–49). Oxford University Press.

Radick, G. (2012). Should "heredity" and "inheritance" be biological terms? William Bateson's change of mind as a historical and philosophical problem. *Philosophy of Science*, *79*(5), 714–724.

Reines, F. M., & Prinz, J. (2009). Reviving Whorf: The return of linguistic relativity. *Philosophy Compass*, *4*(6), 1022–1032.

Rajaram, S. (1993). Remembering and knowing: Two means of access to the personal past. *Memory & Cognition*, *21*(1), 89–102.

Rajsic, J., Wilson, D. E., & Pratt, J. (2015). Confirmation bias in visual search. *Journal of Experimental Psychology: Human Perception and Performance*, *41*(5), 1353–1364.

Reese, H. E., McNally, R. J., & Wilhelm, S. (2010). Facial asymmetry detection in patients with body dysmorphic disorder. *Behaviour Research and Therapy*, *48*(9), 936–940.

Rey, G. (1983). Concepts and stereotypes. *Cognition*, *15*(1–3), 237–262.

Rey, G. (1985). Concepts and conceptions: A reply to Smith, Medin, and Rips. *Cognition*, *19*(3), 297–303.

Rivera, R., & Borda, T. (2001). The etiology of body dysmorphic disorder. *Psychiatric Annals*, *31*(9), 559–563.

Robins, S. K. (2016a). Misremembering. *Philosophical Psychology*, *29*(3), 432–447.

Robins, S. K. (2016b). Optogenetics and the mechanism of false memory. *Synthese*, *193*(5), 1561–1583.

Robins, S. (2016c). Representing the past: Memory traces and the causal theory of memory. *Philosophical Studies*, *173*(11), 2993–3013.

Robins, S. (2017). Memory traces. In Bernecker, S., & Michaelian, K. (eds.), *The Routledge handbook of philosophy of memory* (pp. 76–87). Routledge.

Robins, S., & Schulz, A. (forthcoming). Episodic memory, simulated future planning, and their evolution.

Roediger, H. L., & McDermott, K. B. (1995). Creating false memories: Remembering words not presented in lists. *Journal of Experimental Psychology: Learning, Memory, and Cognition*, *21*(4), 803–814.

Rosch, E. (1975). Cognitive representations of semantic categories. *Journal of Experimental Psychology: General*, *104*(3), 192–233.

Rosch, E. (1978). Principles of categorization. In E. Rosch and B. B. Lloyd (eds.), *Cognition and categorization* (pp. 28–49). Erlbaum.

Rosch E., & Mervis C. B. (1975). Family resemblances: Studies in the internal structure of categories. *Cognitive Psychology*, *7*(4), 573–605.

Rosch, E., Simpson, C., & Miller, R. S. (1976). Structural bases of typicality effects. *Journal of Experimental Psychology: Human Perception and Performance*, *2*(4), 491–502.

Rosen, D. A. (1975). An argument for the logical notion of a memory trace. *Philosophy of Science*, *42*(1), 1–10.

Rosen, J. (1995). The nature of body dysmorphic disorder and treatment with cognitive behavior therapy. *Cognitive and Behavioural Practice*, *2*(1), 143–166.

Ross, L. N. (2020). Causal concepts in biology: How pathways differ from mechanisms and why it matters. *The British Journal for the Philosophy of Science*, *72*(1), 131–158.

Rupert, R. D. (2013). Memory, natural kinds, and cognitive extension; or, Martians don't remember, and cognitive science Is not about cognition. *Review of Philosophy and Psychology*, *4*(1), 25–47

Samuels, R. (1998). Evolutionary psychology and the massive modularity thesis. *British Journal for the Philosophy of Science*, *49*(4), 575–602.

Samuels, R. (2002). Innateness in cognitive science. *Trends in Cognitive Sciences*, *8*(3), 136–141.

Samuels, R. (2007). Is innateness a confused notion? In P. Carruthers, S. Laurence, and S. Stich (eds.), *The innate mind: Foundations and the future* (pp. 17–36). Oxford University Press.

Samuels, R. (2009a). Delusion as a natural kind. In M. Broome and L. Bortolotti (eds.), *Psychiatry as cognitive neuroscience: Philosophical perspectives* (pp. 49–79). Oxford University Press.

Samuels, R. (2009b). The magical number two, plus or minus: Dual process theory as a theory of cognitive kinds. In J. St. B. Evans and K. E. Frankish (eds.), *In two minds: Dual processes and beyond* (pp. 129–146). Oxford University Press.

Samuels, R., & Ferreira, M. (2010). Why don't concepts constitute a natural kind? *Behavioral and Brain Sciences*, *33*(2–3), 222–223.

Sander, D., Grafman, J., & Zalla, T. (2003). The human amygdala: An evolved system for relevance detection. *Reviews in the Neurosciences*, *14*(4), 303–316.

Saxe, R. (2010). The right temporo-parietal junction: A specific brain region for thinking about thoughts. In A. Leslie and T. German (eds.), *Handbook of theory of mind*. Taylor & Francis.

Saxe, R., Brett, M., & Kanwisher, N. (2006). Divide and conquer: A defense of functional localizers. *NeuroImage*, *30*(4), 1088–1096.

Schacter, D. L. (1996). *Searching for memory: The brain, the mind, and the past*. Basic Books.

Schacter, D. L. (2007). Memory: Delineating the core. In *Science of memory: Concepts* (pp. 23–28). Oxford University Press.

Schacter, D. L., & Addis, D. R. (2009). On the nature of medial temporal lobe contributions to the constructive simulation of future events. *Philosophical Transactions of the Royal Society B: Biological Sciences*, *364*(1521), 1245–1253.

Schacter, D. L., Addis, D. R., & Buckner, R. L. (2007). The prospective brain: Remembering the past to imagine the future. *Nature Reviews Neuroscience*, *8*(9), 657–661.

Schacter, D. L., Addis, D. R., Hassabis, D., Martin, V. C., Spreng, R. N., & Szpunar, K. K. (2012). The future of memory: Remembering, imagining, and the brain. *Neuron*, *76*(4), 677–694.

Schaffer, J. (2003). Is there a fundamental level? *Nous*, *37*(3), 498–517.

Schaffner, K. F. (1998). Genes, behavior, and developmental emergentism: One process, indivisible? *Philosophy of Science*, *65*(2), 209–252.

Schectman, M. (1994). The truth about memory. *Philosophical Psychology*, *7*(1), 3–18.

Scholz, J., Triantafyllou, C., Whitfield-Gabrieli, S., Brown, E. N., & Saxe, R. (2009). Distinct regions of right temporo-parietal junction are selective for theory of mind and exogenous attention. *PloS ONE*, *4*(3), e4869.

Sera, M. D., Elieff, C., Forbes, J., Burch, M. C., & Dubois, D. P. (2002). When language affects cognition and when it does not: An analysis of grammatical gender and classification. *Journal of Experimental Psychology: General*, *131*(3), 377–397.

Shagrir, O., & Bechtel, W. (2017). Marr's computational level and delineating phenomena. In D. M. Kaplan (ed.), *Explanation and integration in mind and brain science* (pp. 190–214). Oxford University Press.

Shapiro, L. A. (2004). *The mind incarnate*. MIT Press.

Shea, N. (2003). Does externalism entail the anomalism of the mental? *Philosophical Quarterly*, *53*(211), 201–213.

Shea, N. (2012a). New thinking, innateness, and inherited representation. *Philosophical Transactions of the Royal Society B: Biological Sciences*, *367*(1599), 2234–2244.

Shea, N. (2012b). Genetic representation explains the cluster of innateness-related properties. *Mind and Language, 27*(4), 466–493.
Shea, N. (2013). Inherited representations are read in development. *British Journal for the Philosophy of Science, 64*(1), 1–31.
Shea, N. (2018). *Representation in cognitive science.* Oxford: Oxford University Press.
Simon, H. (1969/1996). *The sciences of the artificial* (3rd edition). MIT Press.
Slobin, D. (1996). From "thought and language" to "thinking for speaking." In J. J. Gumperz and S. C. Levinson (eds.), *Rethinking linguistic relativity* (pp. 70–96). Cambridge University Press.
Slobin, D. I. (2003). Language and thought online: Cognitive consequences of linguistic relativity. In D. Gentner and S. Goldin-Meadow (eds.), *Language in mind: Advances in the study of language and thought* (pp. 157–191). MIT Press.
Sloman, S., Lombrozo, T., & Malt, B. (2007). Ontological commitments and domain specific categorisation. In M. J. Robert (ed.) *Integrating the mind: Domain general versus domain specific processes in higher cognition* (pp. 105–129). Taylor & Francis.
Smith, E. E., & Medin, D. L. (1981). *Categories and concepts.* Harvard University Press.
Smith, E. E., Medin, D. L., & Rips, L. J. (1984). A psychological approach to concepts: Comments on Rey's "concepts and stereotypes." *Cognition, 17*(3), 265–274.
Sober, E. (1999). The multiple realizability argument against reductionism. *Philosophy of Science, 66*(4), 542–564.
Solomon, G. E. A. (1990). Psychology of novice and expert wine talk. *American Journal of Psychology, 103*(4), 495–517.
Spelke, E. (2003). What makes us smart? Core knowledge and natural language. In D. Gentner and S. Goldin-Meadow (eds.), *Language in mind* (pp. 277–311). MIT Press.
Spelke, E. S., & Newport, E. L. (1998). Nativism, empiricism, and the development of knowledge. In R. M. Lerner (ed.) *Handbook of child psychology: Volume 1: Theoretical models of human development* (pp. 275–340). Wiley.
Sperber, D. (1994). The modularity of thought and the epidemiology of representations. In L. A. Hirschfeld and S. A. Gelman(eds.) *Mapping the mind: Domain specificity in cognition and culture* (pp. 39–67). Cambridge University Press.
Spunt, R. P., & Adolphs, R. (2017). A new look at domain specificity: Insights from social neuroscience. *Nature Reviews Neuroscience, 18*(9), 559–567.
Spunt, R. P., & Lieberman, M. D. (2012). Dissociating modality-specific and supramodal neural systems for action understanding. *Journal of Neuroscience, 32*(10), 3575–3583.
Stanley, M. L., Henne, P., Yang, B. W., & De Brigard, F. (2020). Resistance to position change, motivated reasoning, and polarization. *Political Behavior, 42*(3), 891–913.
Stanovich, K. (2010) *Rationality and the reflective mind.* Oxford University Press.
Stanovich, K. E. (2016) The comprehensive assessment of rational thinking, *Educational Psychologist, 51*(1), 23–34

Stanovich, K. E., Toplak, M. E., & West, R. F. (2008). The development of rational thought: A taxonomy of heuristics and biases. In R. V. Kail (ed.), *Advances in child development and behavior: Vol. 36* (pp. 251–285). Elsevier Academic Press.

Stanovich, K. E., Toplak, M. E., & West, R. F. (2020). Intelligence and rationality. In R. J. Sternberg (ed.), *Cambridge handbook of intelligence* (2nd edition) (pp. 1106–1139). Cambridge University Press

Stanovich, K. E., West, R. F., & Toplak, M. E. (2013). Myside bias, rational thinking, and intelligence. *Current Directions in Psychological Science*, 22(4), 259–264.

Stein, E. (1996). *Without good reason: The rationality debate in philosophy and cognitive science*. Clarendon Press.

Stein, E. (1997). Can we be justified in believing that humans are irrational? *Philosophy and Phenomenological Research*, 57(3), 545–565.

Sterelny, K. (2003). *Thought in a hostile world: The evolution of human cognition*. Blackwell.

Sterelny, K., & Griffiths, P.E. (1999). *Sex and death: An introduction to the philosophy of biology*. University of Chicago Press.

Sternberg R. J. (2009) Domain-generality versus domain-specificity of creativity. In P. Meusburger, J. Funke, and E. Wunder (eds.) *Milieus of creativity: Knowledge and space, Vol. 2* (pp. 25–38). Springer.

Stinson, C. (2016). Mechanisms in psychology: Ripping nature at its seams. *Synthese*, 193(5), 1585–1614.

Stich, S. P. (1975). Introduction: The idea of innateness. In S. P. Stich (ed.), *Innate ideas* (pp. 1–22). University of California Press.

Stich, S. P. (1985). *From folk psychology to cognitive science: The case against belief*. MIT Press.

Suddendorf, T., & Corballis, M. C. (1997). Mental time travel and the evolution of the human mind. *Genetic, Social, and General Psychology Monographs*, 123(2), 133–167.

Suddendorf, T., & Corballis, M. C. (2007). The evolution of foresight: What is mental time travel, and is it unique to humans? *Behavioral and Brain Sciences*, 30(3), 299–313.

Suddendorf, T., Addis, D. R., & Corballis, M. C. (2009). Mental time travel and the shaping of the human mind. *Philosophical Transactions of the Royal Society B: Biological Sciences*, 364(1521), 1317–1324.

Suppes, P., Han, B., Epelboim, J., & Lu, I. Z. L. (1999). Invariance of brainwave representations of simple visual images and their names. *Proceedings of the National Academy of Sciences*, 96(25), 14658–14663.

Szpunar, K. (2010). Episodic future thought: An emerging concept. *Perspectives on Psychological Science*, 5(2), 142–162.

Taber, C. S., & Lodge, M. (2006). Motivated skepticism in the evaluation of political beliefs. *American Journal of Political Science*, 50(3), 755–769.

Thompson, R. F. (1991). Are memory traces localized or distributed? *Neuropsychologia*, 29(6), 571–582.

Thompson, R. F. (2007). Persistence: Discrepancies between behaviors and brains. In H. L. Roediger, Y. Dudai, and S. M. Fitzpatrick (eds.), *Science of memory: Concepts* (pp. 199–201). Oxford University Press.

Toh, W., Rossell, S., & Castle, D. (2009). Body dysmorphic disorder: A review of current nosological issues and associated cognitive deficits. *Current Psychiatry Reviews*, *5*(4), 261–270.

Toh, W. L., Castle, D. J., Mountjoy, R. L., Buchanan, B., Farhall, J., & Rossell, S. L. (2017). Insight in body dysmorphic disorder (BDD) relative to obsessive-compulsive disorder (OCD) and psychotic disorders: Revisiting this issue in light of DSM-5. *Comprehensive psychiatry*, *77*, 100–108.

Toplak, M. E., West, R. F., & Stanovich, K. E. (2011). The cognitive reflection test as a predictor of performance on heuristics-and-biases tasks. *Memory and Cognition*, *39*(7), 1275–1289.

Toplak, M. E., West, R. F., & Stanovich, K. E. (2013). Assessing the development of rationality. In H. Markovits (ed.), *The developmental psychology of reasoning and decision-making* (pp. 7–35). Psychological Press.

Tsou, J. Y. (2013). Depression and suicide are natural kinds: Implications for physician-assisted suicide. *International Journal of Law and Psychiatry*, *36*(5–6), 461–470.

Tsou, J. Y. (2016). Natural kinds, psychiatric classification and the history of the DSM. *History of Psychiatry*, *27*(4), 406–424.

Tulving, E. (1972). Episodic and semantic memory. In E. Tulving and W. Donaldson (eds.), *Organization of memory*, (pp. 381–403). Academic Press.

Tulving, E. (1985). Memory and consciousness. *Canadian Psychology/Psychologie Canadienne*, *26*(1), 1–12.

Tulving, E. (2002). Episodic memory: From mind to brain. *Annual Review of Psychology*, *53*(1), 1–25.

Tulving, E. (2007). Coding and representation: Searching for a home in the brain. In H. L. Roediger, Y. Dudai, and S. M. Fitzpatrick (eds.), *Science of memory: Concepts* (pp. 65–68). Oxford University Press.

Tulving, E., Schacter, D. L., McLachlan, D. R., & Moscovitch, M. (1988). Priming of semantic autobiographical knowledge: A case study of retrograde amnesia. *Brain and Cognition*, *8*(1), 3–20.

Tversky, A., & Kahneman, D. (1974). Judgment under uncertainty: Heuristics and biases. *Science: New Series*, *185*(4157), 1124–1131.

Tyler, L. K., & Moss, H. E. (2001). Towards a distributed account of conceptual knowledge. *Trends in Cognitive Sciences*, *5*(6), 244–252.

Van Brakel, J. (1995). Consciousness is not a natural kind. *Behavioral and Brain Sciences*, *18*(2), 269–270.

van Dam, W. O., van Dijk, M., Bekkering, H., & Rueschemeyer, S. A. (2012). Flexibility in embodied lexical-semantic representations. *Human Brain Mapping*, *33*(10), 2322–2333.

van Dam, W. O., Brazil, I. A., Bekkering, H., & Rueschemeyer, S. A. (2014). Flexibility in embodied language processing: Context effects in lexical access. *Topics in Cognitive Science*, *6*(3), 407–424.

Veale, D. (2004). Advances in a cognitive behavioural model of body dysmorphic disorder. *Body Image*, *1*(1), 113–125.

Veale, D., & Riley, S. (2001). Mirror, mirror on the wall, who is the ugliest of them all? The psychopathology of mirror gazing in body dysmorphic disorder. *Behaviour Research and Therapy*, 39(12), 1381–1393.

Venn, J. (1889/1907). *The principles of empirical or inductive logic* (2nd edition). Macmillan.

Waismeyer, A., Meltzoff, A. N., & Gopnik, A. (2015). Causal learning from probabilistic events in 24-month-olds: An action measure. *Developmental Science*, *18*(1), 175–182.

Wakefield, J. C. (1992). The concept of mental disorder: On the boundary between biological facts and social values. *American Psychologist*, *47*(3), 373–388.

Wason, P.C. (1960). On the failure to eliminate hypotheses in a conceptual task. *Quarterly Journal of Experimental Psychology*, *12*(3), 129–140.

Waxman, S. R., & Gelman, S. A. (2010). Different kinds of concepts and different kinds of words: What words do for human cognition. In D. Mareschal, P. C. Quinn, and S. E. G. Lea (eds.), *Oxford series in developmental cognitive neuroscience: The making of human concepts* (pp. 99–129). Oxford University Press.

Weinberg, J. M., & Mallon, R. (2008). Living with innateness (and environmental dependence too). *Philosophical Psychology*, *21*(3), 415–424.

Weiskopf, D. A. (2007). Concept empiricism and the vehicles of thought. *Journal of Consciousness Studies*, *14*(9–10), 156–183.

Weiskopf, D. A. (2009). The plurality of concepts. *Synthese*, *169*(1), 145–173.

Weiskopf, D. A. (2011). The functional unity of special science kinds. *British Journal for the Philosophy of Science*, *62*(2), 233–258.

Weiskopf, D. A. (2017a). An ideal disorder? Autism as a psychiatric kind. *Philosophical Explorations*, *20*(2), 175–190.

Weiskopf, D. A. (2017b). The explanatory autonomy of cognitive models. In D. M. Kaplan (ed.), *Explanation and integration in mind and brain science* (pp. 44–69). Oxford University Press.

Werker, J. F., & Tees, R. C. (1984). Cross-language speech perception: Evidence for perceptual reorganization during the first year of life. *Infant Behavior and Development*, *7*(1), 49–63.

Werker, J. F., & Tees, R. C. (2005). Speech perception as a window for understanding plasticity and commitment in language systems of the brain. *Developmental Psychobiology*, *46*(3), 233–251.

Werning, M., & Cheng, S. (2017). Taxonomy and unity of memory. In S. Bernecker and K. Michaelian (eds.), *The Routledge handbook of philosophy of memory* (pp. 7–20). Routledge.

Whewell, W. (1840/1847). *The philosophy of the inductive sciences* (2nd edition). John W. Parker.

Wierzbicka, A. (2007). Is "remember" a universal human concept? "Memory" and culture. In M. Amberber (ed.), *Memory in language and thought* (pp. 13–39). John Benjamins Publishing Company.

Wilson-Mendenhall, C. D., Simmons, W. K., Martin, A., & Barsalou, L. W. (2013). Contextual processing of abstract concepts reveals neural representations of non-linguistic semantic content. *Journal of Cognitive Neuroscience*, *25*(6), 920–935.

Wimsatt, W. C. (1999). Generativity, entrenchment, innateness, and evolution: Philosophy, evolutionary biology, and conceptual foundations of science. In V. G. Hardcastle (ed.), *Where biology meets psychology: Philosophical essays* (pp. 139–179). MIT Press.

Wimsatt, W. C. (2007). *Re-engineering philosophy for limited beings: Piecewise approximations to reality*. Harvard University Press.

Winawer, J., Witthoft, N., Frank, M. C., Wu, L., Wade, A. R., & Boroditsky, L. (2007). Russian blues reveal effects of language on color discrimination. *Proceedings of the National Academy of Science*, *104*(19), 7780–7785.

Wiser, M. (1988). The differentiation of heat and temperature: History of science and novice-expert shift. In S. Strauss (ed.), *Ontogeny, phylogeny, and historical development*. Ablex.

Wiser, M., & Carey, S. (1983). When heat and temperature were one. In D. Gentner and A. L. Stevens (eds.), *Mental models* (pp 267–297). Lawrence Erlbaum.

Wolfe, C. R., & Britt, M. A. (2008). The locus of the myside bias in written argumentation. *Thinking & Reasoning*, *14*(1), 1–27.

Wolff, P., & Holmes, K. J. (2011). Linguistic relativity. *WIREs Cognitive Science*, *2*(3), 253–265.

Wood, T., & Porter, E. (2019). The elusive backfire effect: Mass attitudes' steadfast factual adherence. *Political Behavior*, *41*(1), 135–163.

Woodward, J. (2017). Explanation in neurobiology: An interventionist perspective. In D. M. Kaplan (ed.), *Explanation and integration in mind and brain science* (pp. 70–100). Oxford University Press.

Woodward, T. S., Moritz, S., Cuttler, C., & Whitman, J. C. (2006). The contribution of a cognitive bias against disconfirmatory evidence (BADE) to delusions in schizophrenia. *Journal of Clinical and Experimental Neuropsychology*, *28*(4), 605–617.

Yee, E., & Thompson-Schill, S. L. (2016). Putting concepts into context. *Psychonomic Bulletin and Review*, *23*(4), 1015–1027.

Ylikoski, P., & Pöyhönen, S. (2015). Addiction-as-kind hypothesis. *International Journal of Alcohol and Drug Research*, *4*(1), 21–25.

Young, L., & Saxe, R. (2009). Innocent intentions: A correlation between forgiveness for accidental harm and neural activity. *Neuropsychologia*, *47*(10), 2065–2072.

Yovel, G., & Kanwisher, N. (2004) Face perception: Domain specific not process specific. *Neuron*, *44*(5), 889–898.

Zachar, P. (2000). Psychiatric disorders are not natural kinds. *Philosophy, Psychiatry, and Psychology*, *7*(3) 167–182.

Index

accommodation thesis, 5
acquisition
   concept, 56
anchoring and adjusting, 193
anti-realism, 11, 30
anti-reductionism, 2, 18, 25
autonoetic phenomenology, 130–132, 155

backfire effect, 203, 204
belief polarization, 203
bias, 186, 187, 192, 208, 235
   affect substitution, 188
   anchoring effect, 188
   belief bias, 188
   bias against disconfirmatory evidence, 204, 205, 224, 225, 227, 229
   cognitive, xviii, 181–209, 222, 223, 225, 229
   confirmation, xviii, 184, 188, 202, 204, 205, 207
   conjunction error, 188
   denominator neglect, 188
   disconfirmation, 201, 204
   focal, 195
   framing effect, 188, 193
   hindsight bias, 188
   impulsively associative thinking, 188
   myside, 184, 199, 202, 205, 207, 209, 211, 234
   outcome bias, 188
   positive response, 205
   representativeness, 188
   simplification, 189, 192
   valuation, 193
   verification, 192
Boyd, Richard, 5, 6, 89, 90, 92, 94
Brown Assessment of Beliefs Scale, 219

canalization, 77, 80, 83
Carey, Susan, 62, 65, 80, 82, 85, 111
category, xv, 189
   cluster, 237–39
   clutter, 79

cognitive, xii, xiii, xiv, xv, xvi, xx, 17, 22, 23, 123, 232, 240, 241
   mental, 242
   neural, xii, xiii, xiv, xvi, xx, 22–24, 241
   ontological, xix, 12, 237–239
   psychiatric, 210
   psychological, 17
   scientific, xv, 157
   taxonomic, xi, xii, xiv, 31, 37, 67, 156, 175, 189, 204, 231, 238, 239, 242
causal history, 152, 155, 233, 235, 236
causal model, xix, 194, 196, 205, 210, 215, 221–230
causal network, 222, 230
causal thickets, 26
classical conditioning, 124, 151
classification, 3–6, 8–10, 25, 64, 123, 162, 192, 212
   clinical psychiatry, 229
   natural, 4
   psychiatric, 218
   scientific, 3
closed process invariance, 77
cognitive architecture, xviii, 184, 185, 194–196, 199, 202, 208
cognitive impenetrability, 80, 85
cognitive miserliness, 181–209
cognitive neuroscience, xvi, xix, 37, 39, 52, 109, 232, 239, 240
cognitive psychology, xvi, 22, 36, 46, 52, 54, 58, 61, 71, 106, 155, 204
Cognitive Reflection Test, 197
cognitive science, xiv, 1, 7, 28, 31, 67, 68, 70, 75, 78–81, 97, 98, 100–102, 106, 108, 109, 121, 123, 124, 127, 135, 150, 155, 157, 174, 181–183, 185, 188, 189, 204, 232, 233, 235, 237, 240–242
Comprehensive Assessment of Rational Thinking, 197
computer science, 2, 184

concept, xv, xvi, 29–31, 75, 84, 94–98, 106, 109, 111, 125, 133, 164, 166, 167, 169, 170, 172, 173, 176, 179, 180, 231, 235, 236, 238, 241, 242
    acquisition, xviii, 33, 36, 72, 172
    activation, 34, 36, 37, 41, 62, 67, 94–98
    amodal theory, 34, 45
    atomism, 35, 50
    classical theory, 94–98
    classical theory of, 34
    eliminativism, 67
    empiricism, 52
    functional theory of, 36
    holism, 35, 50, 73
    hybrid theory, 94–98
    individuation, 33–35, 37, 41, 94–98
    intellectualism, 35, 72
    interpretivism, 29
    maximalism, 73
    minimalism, 35, 72, 73
    modal theory, 34, 45, 67, 73, 94–98
    pluralism, 71
    pluralist theory, 94–98
    possession, 34, 36, 37, 47, 62, 67, 70, 94–98
    prototype theory, 34, 46, 63, 67, 69, 73, 94–98
    scientific, 175
    theory theory, 34, 46, 63, 67, 73, 94–98, 106
concepts, 2, 178, 234
conjunction fallacy, 193
connectivity
    effective, 42
    functional, 42
constructivism, 140–143, 146, 147, 154
contaminated mindware, 192
contextualism, xii, xviii
core cognition, 80, 82, 84, 85
critical period, 80, 86, 87, 93
crosscutting, 22, 23, 214

decision theory, 181, 183
default mode network, 137, 139, 154
delusion, 217, 220–223, 225, 226
developmental psychology, 50, 59, 65, 80, 81, 109, 111, 234, 236
Diagnostic and Statistical Manual of Mental Disorders, xviii, 210, 212, 214, 218, 222
disorder, 230
    body dysmorphic, xviii, xix, 204, 210–230, 242
    obsessive compulsive, xix, 221, 222, 224, 226, 210–230
    psychiatric, 211, 212, 222–224, 228–230
    the explanatory criterion, 211
    the value criterion, 211
domain specificity, xvii, 111, 234, 235, 238, 242
    grain problem, 111, 117, 118

dual-process theory, 55, 185
dual-system theory, 55, 185–187, 194
dystonicity, 220, 226

ecological rationality, 206
eliminativism, 140–142, 154, 242
engram, 141, 142, 151, 152
entrenchment, 77
essentialism, 3, 6, 31, 78, 113, 114
etiological
    individuation, 229
etiology
    natural, 66
    social, 66
evolutionary psychology, 97, 100, 117
explanation
    levels of, xvi, 54, 62
extended mind thesis, 233
externalism, xix, 23, 57–58, 60, 62, 64, 73, 97, 156–158
    active, 233

Fodor, Jerry, 70, 97, 102
folk psychology, 58, 61, 64, 67, 134
function, xiii
    cognitive, 2
functionalism, 120
    wide, xvi

globalism, xii

heuristic, 238
    cognitive, xviii, 230, 231
    myside, xviii, 202, 204–207, 209, 223, 232, 235, 238, 242
    positive test strategy, 200
    recognition, 187
    search, 184
history
    causal, xii, 23, 24, 29, 61, 70, 72, 132, 133
homeostasis, 6
homology, 154
    behavioral, 120
    phenotypic, 151

individuation, xiii, xvi
    concept, 69, 73, 97
    contextual, xiv, xviii
    etiological, xii, xvii, xix, 22, 23, 28, 56, 58, 62, 63, 67, 72, 122, 153, 155, 208, 209, 232, 233, 241
informational encapsulation, 80, 85, 86, 97
innateness, xvi–xvii, 16–17, 27, 31, 33, 100, 101, 111, 119, 121, 177, 182, 231–235, 238, 242
inner speech, 173

Index

instrumentalism, 30
intentional stance, 62, 69
internalism, 23, 34, 62, 64, 73, 97

Jumping to Conclusions, 223

kind, xv, iii–viii, 2, 7–11, 25, 26, 30, 31, 67, 71, 74, 75, 101, 141, 158, 171, 182, 183, 192, 194, 198, 208, 209, 211, 230, 234, 237
  abstract, 70
  cluster, 94
  cognitive, xiv, xv, xvi, xvii, xviii, xix, 3–8, 12, 24, 25, 27, 29, 31, 56, 63, 67, 70, 74, 75, 88, 89, 92–94, 97, 99–101, 108, 114, 115, 121, 122, 124, 126, 127, 131, 141–143, 147, 149, 155, 158–161, 174, 177–181, 183–185, 188, 189, 193, 194, 197, 199, 201, 202, 207–209, 214, 229, 231–233, 235, 238, 240, 241
  copied, 236
  etiological, 147, 183
  functional, xvi
  hybrid, 235
  mental, 29
  natural, xv, 3, 5, 89, 90, 94, 125, 126, 180
  neural, xviii, 24, 31, 93, 99, 150, 178, 209, 238, 240
  psychiatric, xix, 211, 214
  psychological, 214
  real, xiv, 7, 10, 11, 25, 28–30, 72, 75, 76, 94, 98, 124, 126, 127, 129, 133, 137, 142, 147, 149, 151, 156, 159, 182, 210, 213, 214, 225, 231, 238, 239
  relational, 213
  simple causal theory of, xv, 6
  social, 29
kinds, 55
Kripke, Saul, xv, 3

label-feedback hypothesis, 176
language-thought
  effect, 162, 164, 167, 169, 171–174, 177, 178, 180
  effects, 159, 164
  hypothesis, 159–161, 164, 165, 168, 170–173, 175, 179
  process, xvii, 159, 161, 162, 174, 177, 178, 180, 231
level
  algorithmic, xvi, 27, 62, 67, 71, 73, 124, 135
  computational, xvi, xvii, 27, 62, 67, 70, 71, 73, 124, 126, 127, 142, 151, 152, 236, 239
  implementational, xvi, 27, 67, 70, 71, 73, 117, 124, 135, 142, 238, 241
linguistic determinism, xviii, 159, 171
linguistic relativity, xviii, 159, 171
localism, xi
localization, 20, 103, 110, 111

many-to-many. *See* crosscutting
mapping, xii
  crosscutting, 19
  many-many. *See* crosscutting
  many-to-many. *See* taxonomy, crosscutting
  many-to-one. *See* multiple realization
  one-to-many. *See* neural reuse
  one-to-one, xi, xii
  structure-function, xi, xii, 18
Marr, David, 27, 63, 236, 239, 241
mechanism, 2, 6, 71, 89, 90, 92, 93, 98, 99, 103, 116, 135, 185, 196, 206, 211, 236, 237
memory, 127, 130, 163, 170, 173, 177, 178, 215, 238, 242
  causal theory, xvii, 131, 155
  declarative, 124, 126, 127
  Deese–Roediger–McDermott paradigm, 134, 135
  episodic, xvii, 12, 19, 123–157, 211, 231, 232, 234, 235, 238
  long-term, 166–169, 172, 173, 176
  non-declarative, 124, 126
  procedural, 124, 131
  semantic, 12, 123, 124, 128, 129, 131, 133, 143, 147, 152
  short-term, 166, 169, 172, 176
  trace, 132, 135, 140–144, 150–153, 155, 157
  working, 19, 176
mental causation, 29
mental time travel, 137, 139
mereological paradox, 68
Mill, John Stuart, xv, 3–4, 75
mind-dependence, 25, 29
mind-independence, 28
mindware gap, 183, 189, 192, 193, 198
modularity, 86, 101, 103, 115
multi-functionality. *See* neural reuse
Multiple Correspondence Analysis, 193
multiple realizability, 17, 27
multiple realization, xii, xiii, 17, 20, 21, 23, 45, 93, 99, 152, 214, 230

nativism, 75, 80, 84
naturalism, xiv, xv, 2–6, 8–10, 25, 31, 157
network
  causal, xv, xvi, 5, 80, 87, 97, 98
neural correlates, xvi, xix, 21, 37, 40, 42, 44, 46, 93, 99, 138, 152, 230, 240
neural reuse, xii, xiii, 19–21, 23, 24, 93, 138, 155
neuroimaging, xi, xiii, xvi, 2, 18–22, 36, 43, 45, 110, 137, 138, 140, 142, 152, 216
neuromodulation, xii, xiii, 21, 22
nominalism, 9, 11

ontology, 1, 3, 5, 8–13, 27, 28, 115, 126, 143–145, 232, 237
  cognitive, xi, xiv, xv, xix, xx, viii–xiii, 3, 5, 8–13, 19, 33, 72, 210, 231, 237, 238
  four-category, 12
  layer-cake view, 13, 14, 17
  process, 13
  scientific, 27

perceptual illusion, 182, 183, 208
phonological loop, 176
Physical Symbol System, 184
Principle Component Analysis, 193
properties, xvi
prototype, xvi, 35, 56
psychological primitiveness, 77, 89
psychosis, 217
Putnam, Hilary, xv, 3, 233

rationality, 183, 198, 209
  collective, 207
  ecological, 185, 187
  norms of, 181, 182, 208
  standards of, 223
Reading the Mind in the Eyes Test, 217
realism, xiv, xvi, xix, xx, 9, 11, 24–31
  rainforest, xiv
rebound effect, 203
reduction, 31, 45, 69, 71, 178, 240
  type-type, 18
reductionism, xv, 2, 13–24, 37, 232
response-dependence, 29, 35, 69, 73
response-independence, 35
revisionism, xii, 242

Sapir-Whorf hypothesis, 158
scene construction, 138
schizophrenia, 218, 223, 224
self-reflexivity, 25, 30
semantic dementia, 41
Simple Coordination Thesis, 239
source-monitoring, 148, 154
Spelke, Elizabeth, 80, 111
supervenience, 21
syntonicity, 220, 226

taxonomy, xi, xv, 7–9, 22, 25, 27, 57, 64, 69, 123, 153, 156, 168, 169, 174, 189, 193, 233, 234, 236
  biological, 239
  cognitive, xi, xii, xix, 236
  crosscutting, xiii, xiv, xx
  folk, 64
  heuristics, 194, 198, 208
  heuristics and biases, 192
  neural, xi, xii, xix
  neuroscientific, 24
  psychiatric, xix, 210
  scientific, xix, 3, 6, 64, 124
theory of mind, 101, 217, 222, 224, 225, 227, 229
top-down modulation, 43
triggerability, 235
triggering, 77, 80–81
Tulving, Endel, 128, 129, 139, 142, 143

Venn, John, xv, 4, 6–7
verbal interference, 161, 169, 172, 176, 177
vervet monkey, xvii, 27, 104–105, 107, 108
vitalism, 71

Whewell, William, xv, 3–4, 239
Whorfianism, 165, *See* Sapir-Whorf hypothesis

For EU product safety concerns, contact us at Calle de José Abascal, 56–1°, 28003 Madrid, Spain or eugpsr@cambridge.org.

www.ingramcontent.com/pod-product-compliance
Lightning Source LLC
LaVergne TN
LVHW020342260326
834688LV00045B/1482